Using Risk Analysis for Flood Protection Assessment

Martina Zeleňáková · Lenka Zvijáková

Using Risk Analysis for Flood Protection Assessment

Springer

Martina Zeleňáková
Institute of Environmental Engineering
Technical University of Košice
Košice
Slovakia

Lenka Zvijáková
Institute of Environmental Engineering
Technical University of Košice
Košice
Slovakia

ISBN 978-3-319-52149-7 ISBN 978-3-319-52150-3 (eBook)
DOI 10.1007/978-3-319-52150-3

Library of Congress Control Number: 2016963782

Printed on acid-free paper

This Springer imprint is published by Springer Nature
The registered company is Springer International Publishing AG
The registered company address is: Gewerbestrasse 11, 6330 Cham, Switzerland

Preface

Environmental impact assessment (EIA) is an important process that, prior to approval of the investment plan, can provide a detailed examination of the likely and foreseeable impacts of proposed construction activity on the environment. The objective of this book is to develop a specific methodology for the analysis and evaluation of the environmental impacts of selected constructions, namely flood protection structures, using risk analysis methods. Experience in applying this methodology designed for the process of environmental impact assessment also drives considerations for further improvements or more effective implementation and performance of this process. This book looks into the benefits of using risk analysis techniques to evaluate flood protection structures. In doing so, the results of the environmental impact assessment of selected planned flood protection projects are examined.

The book proposes a methodology for the assessment of environmental impact in water management, and it investigates flood mitigation measures with the aim of selecting the best option for the decision process. This methodology should streamline the environmental impact assessment process applied to constructions in the field of the water management. Moreover, the outcome should lead to the selection of future activities entailing minimum risk to the environment. Comparison of alternatives and designation of the optimal variant are implemented based on selected criteria that objectively describe the characteristic lines of the planned alternatives of activity and their impact on the environment.

Specifically, multiparametric risk analysis is used in the proposed method for environmental impact assessment of flood protection projects. This risk analysis method is intended not only to increase the clarity and precision of the evaluation process, but also to align it with the requirements of the environmental impact assessment system of the European Union. This modification should improve the reliability of the environmental impact assessment, and could moreover also be applied to other infrastructure projects. The designed project in Kružlov, Slovakia is used as a case study to clarify and exemplify the methodology and techniques.

This book reviews the literature of EIA and risk analysis and their interconnection (Chap. 1). A proposed methodology for EIA of selected proposed activities based on risk analysis is described in Chap. 2. Chapter 3 reports the results of research based on the application of the proposed methodology of EIA of flood mitigation measures in Kružlov village (north Slovakia). The conclusions of the research, plus both the theoretical and practical benefits of the book as a tool for decision support and the promotion of sustainable development, are treated in Chap. 4. This final chapter also presents suggestions or recommendations for further research in the field of the methodology of the EIA process.

Košice, Slovakia

Martina Zeleňáková
Lenka Zvijáková

Acknowledgements

The authors would like to thank the reviewers namely Prof. h.c. Prof. Ing. Milan Majerník, Ph.D., Professor at Department of Management at Faculty of Business Economy with seat in Košice, University of Economics in Bratislava, Prof. Katarína Pavličková, Associated Professor at Department of Landscape Management at Faculty of Natural Science, Comenius University of Bratislava and Eng. Milan Luciak, working at Department of Environmental Assessment at Ministry of Environment of the Slovak Republic for their constructive comments.

This work has been supported by the Scientific and Educational Grant Agency of Ministry of Education of the Slovak Republic under project VEGA 1/0609/14.

The authors would like to thank also the publisher—Springer Nature for providing the opportunity for this publication.

Contents

Abbreviations

ALARP	As low as reasonably practicable
AS/NZS	Australian/New Zealand Standard
B.C.	Before Christ
CEAM	Cumulative effects assessment and management
DA	Decision analysis
EA	Environmental assessment
ECOTOC	European Centre for Ecotoxicology and Toxicology of Chemicals
EES	Environmental evaluation system
EIA	Environmental impact assessment
EIS	Environmental impact statements
EU	European Union
EUR	Euro
FAO	Food and Agriculture Organization
FPS	Flood protection structures
GIS	Geographic information system
HIA	Health impact assessment
I&APs	Interested and affected parties
IAIA	International Association for Impact Assessment
IAP2	International Association for Public Participation
IAPA	Impact assessment and project appraisal
IEM	Integrated environmental management
IFC	International Finance Corporation
IMP3	IMProving the IMPlementation of Environmental IMPact Assessment
IUCN	International Union for Conservation of Nature
MAUM	Multi-attribute utility measurement
MCDM	Multi-criteria decision making
MCDM	Multi-criteria discrete model
MoE	Ministry of Environment
MPOM	Multi-purpose optimization model
NEPA	National Environmental Policy Act

NGOs	Non-government organizations
OECD	Organisation of Economic Cooperation and Development
PADC	Project appraisal for development control
PRA	Pest risk analysis
RA	Risk analysis
RIA	Regulatory impact assessment
SA	Sustainability assessment
SEA	Strategic environmental assessment
SHMI	Slovak Hydrometeorological Institute
SIA	Social impact assessment
SWOT	Strengths, weaknesses, opportunities and threats
TIA	Transboundary impact assessment
TIEQ	Total indicator of environmental quality
TSES	Territorial system of ecological stability
UMRA	Universal matrix of risk analysis
UNEP	United Nations Environment Programme
USA	United States of America
USEPA	United States Environmental Protection Agency
V4	Visegrad Four-4
WHO	World Health Organization

Chapter 1
Environmental Impact Assessment—State of the Art

Environmental impact assessment (EIA) is now 44 years old (beginning on 1 January 1970 when President Richard Nixon signed the National Environmental Policy Act in the USA). EIA is a systematic approach to identifying and evaluating positive and negative impacts on components of the environment that may arise from the implementation of infrastructure projects or policies (Petts 1999; Wang et al. 2006; Gilbuena et al. 2013). EIA is a mandatory process before approval of infrastructure projects with significant impacts on the environment (Tamura et al. 1994), such as roads (Zhou and Sheate 2011), water supply systems (Al-agha and Mortaja 2005) and flood protection constructions (Ludwig et al. 1995). Flood protection structures (FPS) have been created throughout the centuries to mitigate flood damage (Poulard et al. 2010; Gilbuena et al. 2013).

In Slovakia, through the National Council of the Slovak Republic's Law No. 24/2006 on the environmental impact assessment of proposed activities, the screening process of planned FPS was made mandatory. The most commonly-used EIA methods are essentially descriptive and qualitative (e.g. Department of Public Works and Highways 1998; City Office of Navotas 2009; Gilbuena et al. 2013). Some of them (especially *ad hoc* and simple checklist methods) are analysed by Lohani et al. (1997).

Ad hoc methods are so called because they are not structured approaches which typically rely on the "experience, training and intuition" of the assessors. The disadvantage of these methods is that they often lack information about the environment. They present information regarding the impacts without emphasizing their magnitude or significance. This type of assessment is also non-replicable, which complicates the review of any EIA conclusions (Lohani et al. 1997).

In contrast, the simple checklist method is structured, elaborate, well organised and more systematic. It usually consists of a series of environmental parameters (or potential impacts) that are compared with assessment criteria (Barthwal 2002; Lohani et al. 1997). One problem with the checklist method is that it often

M. Zeleňáková and L. Zvijáková, *Using Risk Analysis for Flood Protection Assessment*, DOI 10.1007/978-3-319-52150-3_1

overlooks the fact that environmental impacts have spatial and cumulative effects (Munier 2004). This method is also insufficient when it is necessary to estimate and interpret the rates of impacts (Lohani et al. 1997), which improves the transparency of the EIA process (Gilbuena et al. 2013). According to Lexer et al. (2006), MoE and SEA (2012) and Zvijáková (2013), the way to improve the EIA system in the Slovak Republic is to develop methodology or guidance on applying risk analysis in EIA to produce better transparency and preserve the impartiality of the EIA process, the result of which should lead to the choice of future activities associated with minimum risk to the environment.

This chapter begins with a review of literature on the following topics: environmental impact assessment, both state-of-the-art worldwide and in Slovakia (Sect. 1.1); risk analysis and its importance in environmental section (Sect. 1.2). Section 1.3 highlights the implications from the literature and develops the conceptual framework for the study.

1.1 Environmental Impact Assessment

The emergence of EIA as a key component of environmental management since 1970 has coincided with the increasing recognition of the nature, scale and implications of environmental change brought about by human actions. During that time, EIA has developed and changed, influenced by the changing needs of decision-makers and the decision-making process, and by the experience of practice (Morgan 1998). Nowadays it is more important than ever to scrutinize decisions that significantly affect people and communities, and the systems that underlie the natural environment, so it is useful to take stock of the progress made in the field, and to reflect on current and future challenges. Accordingly, this chapter has five parts. The first briefly examines the origins and development of EIA to establish the current extent of EIA usage, the forms of impact assessment that have emerged and the contexts within which EIA is applied. The second part reflects on recent trends in EIA in the areas of theory development, practice and effectiveness, before drawing, in the third part, some broad conclusions about the current state of EIA, and the opportunities that are available to shape the future of EIA. The fourth part reviews the state of the EIA process in the Slovak Republic, from the perspectives of professionally qualified persons as based on a questionnaire survey. The last part presents the methods used in EIA.

1.1.1 Origins and Development of EIA

The National Environmental Policy Act (NEPA) was the first instance of formal incorporation of the impact assessment process in legislation (O'Riordan and Sewell 1981). This act introduced an environmental policy guiding the activities of federal agencies that had the power to affect people, communities and the environment in significant ways, and it came in response to rising scientific concern and popular anxiety about ongoing environmental changes (Ashby 1976).

Following the example of the early adopters (including countries such as Australia, Canada, Eire, Sweden, and New Zealand) (O'Riordan and Sewell 1981; Wood 2003), many other countries have introduced some kind of impact assessment process into their formal procedures or legislation regarding environmental planning or decision-making. In the international arena, the institutionalization of EIA has progressed steadily over the last 15–20 years, gaining momentum from increasing political awareness of the problems arising from climate change, including loss of biodiversity, threats to freshwater sources and water quality, damage to marine areas and other forms of global environmental change (Morgan 2012).

Morgan (2012) gives a concise account of the course of developments since NEPA in the USA to the present day, when some form of EIA has been made mandatory in 191 of the 193 countries of the world. He concludes: "*After 40 years, it seems reasonable to say that EIA is now universally recognised as a key instrument for environmental management, firmly embedded in domestic and international environmental law*". EIA is now used in a wide range of decision-making contexts, including international development and trade policy (Kirkpatrick and George 2006; Cashmore et al. 2009; Pope et al. 2013), as well as disaster anticipation measures and post-disaster recovery (e.g., Srinivas and Nakagawa 2008).

In examining what drives the development and spread of EIA practice, Morgan (2012) stresses the role of international agreements delegating impact assessment, as well as the increasing involvement of international finance providers such as the World Bank, the International Finance Corporation (IFC) and lending institutions applying the Equator Principles, which require EIA on major investment projects.

EIA has in fact become an umbrella term, having given rise to a number of specific forms since the 1970s, including "*social impact assessment* (SIA), *health impact assessment* (HIA) and *strategic environmental assessment* (SEA)". Each of these seems to have developed due to certain dissatisfaction with the way EIA has been implemented. SIA, for example, emerged specifically in the late 1970s and 1980s because EIA, especially in the USA, was seen as having strong biophysical emphasis, apparently neglecting the social impacts (Taylor et al. 2004). More recently HIA has developed as a popular form of impact assessment in response to a sense among many public health professionals that EIA did not adequately address project impacts on community and individual health (National Academy of Sciences 2011). SEA has been strongly promoted as a way of extending impact assessment to higher-level policy and planning decision-making, in reaction to the project orientation of most EIA approaches (Sadler et al. 2011).

A related approach, *sustainability assessment* (SA), has emerged in recent years, its focus being more specifically on sustainability criteria in the assessment of policies, plans or projects. However, sustainable development has a variety of meanings, and as a consequence the process of SA can be viewed in different ways (Pope et al. 2004; Bond and Morrison-Saunders 2011). Moreover, SEA is often justified on the grounds that it promotes sustainable outcomes (Sadler et al. 2011), further blurring the boundaries with SA. Similar problems occur across the impact assessment field, as finer differentiation of the EIA model into named varieties throws up conceptual as well as terminological complexities.

1.1.2 The State of Environmental Impact Assessment

A special edition of *Impact assessment and project appraisal* (IAPA) came out recently seeking to ascertain the status of impact assessment in 2012 with regard to six core branches of practice: "*environmental impact assessment* (Morgan 2012); *strategic environmental assessment* (Fundingsland Tetlow and Hanusch 2012); *policy assessment* (Adelle and Weiland 2012); *social impact assessment* (Esteves et al. 2012); *health impact assessment* (Harris-Roxas et al. 2012); and *sustainability assessment* (Bond et al. 2012)".

Pope et al. (2013) produced a survey of the situation of *impact assessment* in 2013. They made two main points. Firstly, impact assessment practice varies greatly depending on the context of its application, and so EIA may be interpreted quite differently in one jurisdiction compared with another. Secondly, there are various forms which involve special focus. These include "*ecological impact assessment* (Briggs and Hudson 2013; Treweek 1999), *climate change impact assessment*, and *cultural heritage impact assessment*". Other types of impact assessment that have been developed in recent years include "*regulatory impact assessment* (RIA), *human rights impact assessment*, *cultural impact assessment*, *post-disaster impact assessment* (Morgan 2012)".

Impact assessment is thus characterized by diversity in its practice, and by associated ambivalence (Pope et al. 2013). The main challenge for EIA practitioners, however, will be to ensure that all types of impact assessment contribute to the effective assessment of projects, based on shared principles that are accepted by the impact assessment community as a whole, and applied in an integrated and complementary way (Morgan 2012). With this in mind, in the next part of the book, we reflect on recent thinking about EIA, in terms of its theoretical basis, and the extent to which those views are informing ideas about impact-assessment practice and the wider issues of effectiveness.

Retief (2010) presents a conceptual framework bringing together three underlying principles in environmental impact assessment, showing how "*effectiveness* (what we are achieving) is strongly linked to both *theoretical grounding* (what is EIA?) and *quality* (how to conduct EIA)".

1.1.2.1 Theory and EIA

Increasing emphasis has been placed in recent years on development of the theory of EIA, primarily as a consequence of increasing recognition that the theoretical basis of "state-of-the-art" EIA is inadequately developed and detailed (Cashmore 2004). Ortolano and Shepherd (1995) analyse "the nature of EIA, especially the dominance of the technocratic model of impact assessment and the rise of alternative views that recognize the political realities of decision-making". Lawrence (1997) observes that "the conceptual foundation of EIA has received limited attention". Since Lawrence (1997) called for "more coherent impact assessment theory-building, arguing that this theory is essential to further understanding of human activity, the environment, and critical interactions between the two", EIA theoreticians have responded by "discerning different substantive goals of environmental impact assessment as well as different mechanisms by which it may function".

Important sources of thinking about the theoretical basis of EIA have been the various models of planning and decision-making. For example, Lawrence (2000) examined five planning theories: "*rationalism, pragmatism, socio-ecological idealism, political-economic mobilization*, and *communications and collaboration*", while Leknes (2001) uses a simpler triple categorization of decision-making approaches: "the *rational*, *new institutionalist* and *negotiation perspectives*". In contrast, Bartlett and Kurian (1999) adopt a political science perspective, identifying six models which appear to them to be implicit in preceding discussions of EIA in the literature: "the *information processing* model, the *symbolic politics* model, the *political economy* model, the *organizational politics* model, the *pluralist politics* model, the *institutionalist politics* model".

Cashmore (2004) suggested that there are two main interpretations of the role of science in EIA "*EIA as applied science* and *EIA as civic science*" and five distinct models are identified within these paradigms "the *analytical science* model, the *environmental design* model, the *information provision* model, the *participation* model, *environmental governance* model".

Bartlett and Kurian (1999) proposed six models of EIA that are "based primarily on how EIA is perceived to influence decision processes".

From a theoretical point of view, EIA's origin in the information processing or linear rational decision-making models is widely recognized (Morgan 2012; Pope et al. 2013). This characterizes EIA metaphorically as 'knowledge speaking to power', whereby "the information generated by this predictive process contributes (albeit in a variety of ways) to the environmental design of development proposals and the formulation of decisions on whether, and potentially on what terms, development consent should be granted" (Cashmore et al. 2004).

Cashmore et al. (2008) reiterate this approach, and specify four results of EIA's transformative potential: "*developmental outcomes*; *learning outcomes*; *governance outcomes*; and *attitudinal and value changes*, which are by nature longer term and often less tangible" (Cashmore et al. 2009).

1.1.2.2 The Practice of EIA

On the issue of EIA practice, the International Study of the Effectiveness of Environmental Assessment (Sadler 1996) concluded: "*Despite the many methodological and administrative advances in EIA over the past two decades, recent experience in many countries confirms that there is still considerable scope for strengthening the process. Immediate and cost-effective measures could help improve the process in four key areas: scoping, evaluation of significance, review of EA reports, and monitoring and follow-up*".

However, problems with practice persist. For example, a recent report of the state of EIA in the UK based on practitioner opinions identifies problems in four key areas of practice: "*screening*, *scoping and engagement*, *assessment*, and *outcomes and outputs*" (IEMA 2011). An earlier European Union report (European Commission 2009) on the application and effectiveness of the EIA Directive identified a number of areas where improvements in practice are needed, including "*screening*, *scoping*, *consideration of alternatives*, *monitoring*, *public participation* and *EIA quality control*".

In the last twenty years significant contributions have been made to the literature regarding the principal phases of the classic model of EIA procedure:

- *Screening*: e.g., Enserink (2000), Pinho et al. (2010), Rajaram and Das (2011) and Weston (2011).
- *Scoping*: e.g. Mandelik (2005) and Snell and Cowell (2006).
- *Impact prediction*: the widest-ranging field involving specific techniques covering various environmental sectors and forms of human activity, plus broader contributions such as Duinker and Greig (2007) on scenario approaches to prediction.
- *Significance*: either more conceptual, e.g. Lawrence (2007a, b); more technical approaches: e.g., Ijäs et al. (2010), Cloquell-Ballester et al. (2007) and Mustow et al. (2005).
- *Monitoring and other aspects of follow-up*: e.g. Marshall et al. (2005) and Morrison-Saunders and Arts (2004).

The consideration of practical issues in impact assessment constitutes a large proportion of the published EIA literature. Several authors, however, point out the lack of correlation between the availability of guidance and good practice; Adelle and Weiland (2012) for example remark that "in some countries a large gap between the policy assessment system and assessment practice exists".

Several special edition papers identify specific procedural issues which continue to produce concern both regarding practice and in the literature more generally. These are: "*cumulative effects*, *public participation*, and *consideration of alternatives*" (Pope et al. 2013). Canter and Ross (2010) draw attention to the patchy nature of practice in *cumulative effects assessment and management* (CEAM), but produce some examples of good practice and propose a sequence six steps as a guide for practitioners. *Public participation* is a firmly-rooted approach in its own right, as demonstrated by the existence of organizations such as the International Association for Public Participation (IAP2) (www.iap2.org), which is represented strongly within the impact assessment community (Stewart and Sinclair 2007). On the other hand, *consideration of alternatives* has generally been considered a weak approach for some time (Steinemann 2001), a view that has recently reappeared in connection with policy analysis (Adelle and Weiland 2012).

1.1.2.3 EIA Effectiveness

The conclusion of the *International Study on the Effectiveness of Environmental Assessment* was that, while EA had made its mark since it was introduced 25 years earlier, it would be necessary to maintain the efforts to improve its performance if it was to make a substantive contribution to the goal of sustainable development (Sadler 1996).

The theme of effectiveness of EIA has been ever present in the literature since then, but as Cashmore et al. (2004) observe, the bulk of that literature addresses procedural issues, with a much smaller proportion concerned with substantive issues. Both are important parts of the overall assessment of effectiveness, but the *procedural aspects* are more amenable to study and analysis, while substantive considerations raise more difficult questions.

The special issue of IAPA (Bond and Pope 2012) updates the last international association for impact assessment (IAIA) overview of the field that was published in 1995 (Vanclay and Bronstein), which was followed in 1996 by the *International Effectiveness Study* (Sadler 1996). In 2006, an update to the *International Study on the Effectiveness of Environmental Assessment* was initiated by the IAIA and the final report from that process is pending (Morgan 2012).

Cashmore et al. (2004) refer to "*the interminable issue of effectiveness*", and "the effectiveness of impact assessment practice certainly continues to receive considerable attention". Adelle and Weiland (2012) point out that effectiveness is evaluated exclusively on the basis of the apparent purpose of each impact assessment and the actual mechanisms with which it is implemented, as Jay et al. (2007) also later agreed. Another well-established idea is that "(t)he greatest contribution of EIA may well be in reducing adverse impacts before proposals reach the decision-making stage" (Wathern 1988). More recent studies also confirm the role of the "*institutional politics*" model, which shows how political institutions become transformed with time as they acquire organizational experience (Sánchez and Morrison-Saunders 2011; Waldeck et al. 2003).

Mutual learning also happens through interactions between regulators and consultants working for project developers (Morrison-Saunders and Bailey 2009). Such learning on the part of decision-makers, at both organizational and individual levels, has also been quoted as an important aspect of EIA effectiveness in the literature (e.g., Fitzpatrick 2006; Jha-Thakur et al. 2009).

Steady streams of evaluations of national EIA systems have been published. These take stock of practical experience and identify and seek solutions for shortcomings, but they also provide feedback on innovative practices, new areas of application of EIA, or new challenges to be recognized and addressed. A deeper investigation into many such national evaluations reveals two key issues (Morgan 2012). First, any evaluation of EIA effectiveness is only meaningful when made in the socio-economic, political and cultural context of the country or countries concerned. This is well presented by the comparison of EIA in Kenya, Tanzania and Rwanda (Marara et al. 2011) in which shortcomings in the EIA system in Rwanda can be attributed to weaker institutional structures, and a comparative lack of local capacity to work with EIA. Similarly, evaluations of EIA in member states of the EU must always be interpreted within the political and institutional context of that grouping, and the overarching framework of the EIA Directive (Wood 2003). Second, views on effectiveness depend on one's understanding of the nature and purpose of EIA, a point made by Elling (2009). It is interesting, for example, to contrast the technical, engineering perspective of Kruopienė et al. (2009), who laments the politicization of EIA in Lithuania, and calls for much stronger recognition for the role of experts in the process, with the characterization of EIA in the Philippines by Bravante and Holden (2009).

Here we have two very different visions of EIA, one clearly rooted in "the *information processing* (rationalist) model, and the other a variant of the *symbolic politics* model", to use Bartlett and Kurian's (1999) terms. Each brings to bear a perspective that influences how the purpose of EIA and its effectiveness, in either procedural or substantive terms, or both, are viewed. It is important that such differences in perspective are recognized and made explicit, if the debate on effectiveness is to move forward in a constructive way (Elling 2009).

If we look again at the political models of EIA suggested by Bartlett and Kurian (1999) (or indeed the various planning or decision-making models), effectiveness can be seen from a number of different, politically oriented perspectives. Has the process opened opportunities for local people to be more involved in decision-making? Have companies become more aware of environmental issues through EIA and modified their practices accordingly to gain competitive advantage? Has change been brought about in government bodies dealing with, say, natural resources, to internalize EIA thinking? Do decision-makers and other stakeholders, understand and use the EIA information provided to them?

Cashmore et al. (2010) take this line of thinking further, using political theory, and especially the notion of politics as the acquisition or exercise of power, to examine effectiveness evaluation and EIA. They demonstrate the degree to which all aspects of EIA as a process are in essence political—including the established methods for evaluating EIA effectiveness which usually involve a limited number of "experts" using criteria agreed within the group. If EIA is political, then there will be a plurality of views about the way the process operates and what it achieves, and that plurality must be recognized in evaluations of effectiveness. That would then allow the evaluator to examine the basis for, and implications of, the differing perspectives, which would in turn inform policymaking (Cashmore et al. 2010). As with other areas of EIA, thinking about effectiveness has moved beyond the mechanistic, process-oriented models to those informed by more recent theoretical perspectives of values, collaborative processes, power relationships, but above all a more thoughtful and considered approach (Morgan 2012).

Bond et al. (2013) present an alternative system for evaluating impact assessment effectiveness based on six categories: "*procedural* (required stages), *substantive* (outcomes), *transactive* (efficiency), *normative* (legislative interpretations), *knowledge and learning* (stakeholders learn from EIA practice) and *pluralism* (different views of what effectiveness means in each category)". The inclusion of the latter category recognizes the possibility that views differ on how EIA works to further the process of change.

According Pope et al. (2013), "governments are particularly focused on *cost effectiveness* in times of recession and are therefore likely to focus most on *transactive* effectiveness". Should EIA fail to satisfy with regard to this measure, its usefulness might be questioned in the future.

1.1.3 Strengths, Weaknesses, Threats and Opportunities of EIA

The synthesis undertaken provides the basis for an overview of the main strengths, weaknesses, opportunities and threats for EIA, building on the already existing status of EIA as defined by Pope et al. (2013), Morgan (2012), and Bond and Pope (2012). The available SWOT appraisal is summarized in Table 1.1.

What comes through very clearly from the state-of-the-art papers on EIA is that the current global recession is a significant threat to practice. Another clear issue that impact-assessment practice needs to accommodate is climate change and, in many jurisdictions, the consideration of climate impacts is required within impact-assessment legislation. While this has already led to the development of a separate stream of climate impact assessment, which interestingly was the subject of a chapter in Vanclay and Bronstein (1995), the EIA, sustainability assessment and SEA papers explicitly acknowledge the need to better accommodate climate impacts (Bond and Pope 2012).

Despite all these weaknesses and threats, the various EIA processes seem to be flourishing. Despite the concerns over the effectiveness of impact assessment, more and more practitioners apply the approaches to the decisions that are made. There is generally agreement on broad process principles, but debates still exist over methodological approaches; this is inevitable given the different theoretical perspectives that exist and, given the diversity of contexts and resources available for funding impact assessment, is not necessarily a bad thing (Bond and Pope 2012).

Table 1.1 Strengths, weaknesses, opportunities and threats analysis for EIA (adapted from Pope et al. 2013)

Strengths
• "The widespread incorporation, particularly of EIA, into legislation and international agreements, and the increasing acceptance of other, supplementary forms of impact assessment" • "The generally good availability of procedural guidance and the value of this for the purposes of both procedural effectiveness and capacity building, notwithstanding that some practice areas are better established than others and that debates continue regarding the appropriate extent and level of prescription of such guidance" • "A strong international body of practitioners including a growing number of theorists, resulting in a continually evolving field" • "Evidence that impact assessment is having effects through different mechanisms, although these may not be direct or immediately apparent" • "Diverse practices that incorporate a range of different perspectives and theoretical bases, although diversity is something of a double-edged sword"
Weaknesses
• "Poor quality of practice and continued capacity issues in many countries" • "Persistent universal areas of weak practice, including consideration of alternatives, meaningful public participation and cumulative-effects assessment and management" • "An ever-expanding range of discrete forms of practice, each with its own literature and body of practitioners, with unclear relationships between all other areas of practice, and in some cases even an ambiguous raison d'etre" • "A lack of integrated consideration of broader sustainability issues within impact assessment, which could be due to factors including increasing specialization and resultant silos within the profession as well as lack of recognition and hence resources allocated to non-regulated forms of EIA"
Opportunities
• "The innovation that it has promoted for all levels and types of development activity, along with the concepts of follow-up and adaptive management that specifically promote taking a flexible (and arguably creative) approach" • "Emerging opportunities for the future of impact assessment include: an increasing focus on climate change, incorporating both mitigation and adaptation concerns and further incorporation of concepts such as systems dynamics, resilience and ecosystem services into impact assessment" • "Deeper reflection upon the state of the art of impact assessment and perhaps confronting some less than comfortable truths about our field and our profession, particularly in the context of current global challenges"
Threats
• "EIA should be integral to project development and design processes, not left to the final legal step before project implementation. This would allow impact assessors to work more constructively with proponents and stakeholders to develop processes that meet the needs of all parties, and in so doing result in projects that are consistent with the environmental and social aspirations of local communities" • "There is also a risk that the spirit of impact assessment becomes neutered through excessive use of checklists, protocols, guidance or standards so that it is reduced to a licensing exercise with negligible discretionary decision-making opportunities" • "Another trend is for responsibility for EIA to be allocated to government agencies whose mandate is development rather than those whose mandate is environmental protection"

1.1.4 *EIA Methods*

The search for the *ideal method* for satisfying all scientific and policy issues related to the EIA process has been ongoing since 1970. Considerations regarding the term *methods* have evolved over 25 years of EIA practice. For example, in the early years methods typically denoted systematic approaches for identifying and integrating impact concerns; hence, they were seen as consisting of interaction matrices, networks, and checklists. Methods now also include decision-analysis approaches for comparing and selecting a proposed action from several alternatives, monitoring for determining the effectiveness of mitigation measures, and techniques for public participation (Canter 1998a). A more recent development is the introduction into EIA of *risk assessment* methods.

Currently a large number of methods are used within environmental impact assessment. The various methods are extremely differentiated and it is therefore difficult to establish a unit classification. EIA methods can also be described as "methodologies", "technologies", "tools" or "models".

The purpose of this section is not to recommend a single method for assessing the impacts of proposed activities on the environment, but to note the different approaches that can be adapted and combined to suit them to the specific project.

1.1.4.1 Types of Methods

A lot of methods (tools) have been utilized over the last decades to meet the different actions required in the conduction of impact studies. The objectives of the different actions vary, as do the usable methods for each. Table 1.2 delineates 22 types of methods arrayed against seven typical study activities (Canter 1999). An *x* denotes that the listed method type is, or may be directly useful for, a given activity. However, the absence of an *x* for any given type of method does not mean that it has no usefulness for the activity; it merely suggests that it may be indirectly related to the activity. The types of methods in Table 1.2 encompass many specific techniques and tools. The methods are listed alphabetically rather than in order of importance or usage. These methods are describe for example in (Canter 1998a) or (Canter 1999).

Guidelines of European Commission (Walker and Johnston 1999) provide information on eight methods and tools that were selected from case studies and literature research. These generally fall into two groups:

"*Scoping and impact identification techniques*"—these identify how and where an indirect or cumulative impact or impact interaction would occur.

"*Evaluation techniques*"—these quantify and predict the magnitude and significance of impacts based on their context and intensity.

During the EIA process usually a combination of techniques is used, or approaches are adopted at different stages of the project. Examples of both categories are set out below, on the Fig. 1.1.

Table 1.2 Synopsis of EIA methods and study activities (adapted from Canter 1999)

Types of methods in EI	Define issues (scoping)	Impact identification	Describe affected environment	Impact prediction	Impact assessment	Decision making	Communication of results
"Analogues (look-alikes) (case studies)"	x	x		x	x		
"Checklists (simple, descriptive, questionnaire)"		x	x				x
"Decision-focused checklists (MCDM; MAUM; DA; scaling, rating, or ranking: weighting)"					x	x	x
"Expert opinion (professional judgment, Delphi, adaptive environmental assessment, simulation modelling)"		x		x	x		
"Expert systems (impact identification, prediction, assessment, decision making)"	x	x	x	x	x	x	
"Laboratory testing and scale models"		x		x			
"Literature reviews"		x		x	x		
"Matrices (simple, stepped, scoring)"	x	x		x	x	x	x
"Monitoring (baseline)"			x		x		
"Monitoring (field studies of analogues)"				x	x		
"Networks (impact trees and chains)"		x	x	x			
"Overlay mapping (GIS)"			x	x	x		x
"Photographs and photomontages"			x	x			
"Qualitative modeling (conceptual)"			x	x			
"Quantitative modeling (media, ecosystem, visual, archaeological, systems analysis)"			x	x			

(continued)

Table 1.2 (continued)

Types of methods in EI	Define issues (scoping)	Impact identification	Describe affected environment	Impact prediction	Impact assessment	Decision making	Communication of results
“Risk assessment”	x	x	x	x	x		
“Scenarios”				x			
“Trend extrapolation”			x	x			

x potential for direct usage of method for listed activity, *MCDM* multi-criteria decision making, *MAUM* multi-attribute utility measurement, *DA* decision analysis, *GIS* geographical information system

The next differentiation of methods in the EIA process is given in Kozová et al. (1996). Table 1.3 presents the recommended methods suitable for impact assessment in the preliminary impact study, environmental impact statement and methods of multi-criteria analysis.

In developed foreign EIA practice, the following models are considered as significant (Říha 2001):

- Multi-criteria methods,
- Conceptual or qualitative models,
- Quantitative models,
- Dynamic and simulation models,
- Models of ecological effectiveness.

The last two categories of models have not achieved significant application in EIA practice so far.

The chosen typology of continuous and discrete multi-criteria decision models leads to this separation of the two main groups:

- Multi-criteria discrete models (MCDM) and
- Multi-purpose continuous optimization models (MPOM).

Another possible resolution is according to the degree of "softness" and "hardness", i.e., according to the degree of completeness and accuracy of input data. In this respect, the distinction is between models of (Říha 2001):

- indeterminate "soft" type (SOFT),
- determinate "hard" type (HARD).

The typology of methods is given in Table 1.4. Each group is represented in practice by more or less well-known models, methodologies, procedures and designs. No unambiguous synonyms are available for the most frequent proposals; nomenclatural problems and confusion often arise. Due to frequent modifications of the methods used, just using one of their names is not normally sufficient.

In the assessment of the impacts of hydraulic structures, the optimization method of replacement value in the form of value equivalents has stimulated attention (Surrogate Worth Trade-off Method), as it is particularly suitable for the range of four-six criteria (Říha 2001).

The character of each model can be quantitative or qualitative.

Qualitative Models

In the literature, these are also known as conceptual or institutional models. They mostly concern (Říha 1995):

- identification of the essential components of the system,
- qualitative identification of the structure of the system,
- design of the flowchart of the system.

The staff of the University of Aberdeen in 1976 proposed a method for qualitative assessment of the potential impacts of industrial investment under the name

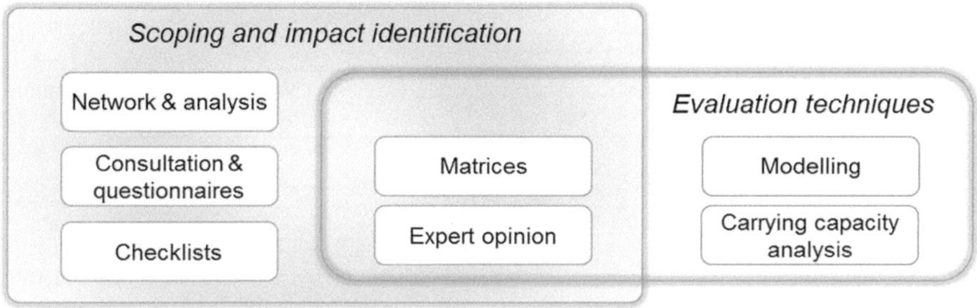

Fig. 1.1 Methods and tools for assessment of indirect and cumulative impact as well as impact interaction (after Walker and Johnston 1999)

Table 1.3 Methods for impact assessment in the preliminary impact study, environmental impact statement and methods of multi-criteria analysis (after Kozová et al. 1996)

Group of methods	Methods
Recommended methods suitable for preliminary impact study	Ad hoc methods Checklists and catalogues of criteria Tables and matrices, expressing the causes and the effects Networks and system diagrams Decision trees Overlay mapping
Recommended methods suitable for environmental impact statement	Prognostic methods • Modeling – Mathematical models – Simulation models – Experimental in situ or laboratory models • Comparative methods
Multi-criteria analysis	Method of utility function TIEQ method Methods for determining weights of criteria • Ranking method • Allocation method • Grading method • Pairwise comparison method • Dual method ALO-FUL

Project appraisal for development control (PADC). It takes the form of a methodological manual giving the recommended instructions for systematic assessment of significant industrial structures and their impact on the environment in the UK.

A more sophisticated qualitative method consists of multi-criterial damage analysis (Concordance Analysis) according to Nijkamp (1980). One of the main features and advantages of this method is the fact that it does not require the

Table 1.4 Typology of multi-dimensional models (after Říha 2001)

Distinction according to completeness of input data	Type of model	
	Multi-criteria discrete MCDM	Multi-purposes optimization MPOM
SOFT	MCDM—I • Expected value methods • Terraced harmony methods • Successive permutations methods • Multidimensional scale methods • Metaplay theory methods • Frequency analysis methods • Method of self-worth • Method of linguistic variable	MPOM—I • Foggy models of "FUZZY" type • Stochastic models and • Indeterminate econometric models
HARD	MCDM—II • Methods of achievable targets • Method of expected value • "Trade-off" analysis • Analysis of consensus • Analysis of entropy • Development analysis • Discriminant analysis and • Method of total indicator of environment quality	MPOM—II • Known models with vector restriction • Target programming models • Hierarchical models • Models of disputed goals of the "Min-Max" type • Models of ideal point • Models of penalization functions • Models of utility function

transformation of planned or expected environmental impacts of a project into financial units. Instead of this it introduces a system of relative importance of individual decisive criteria.

Quantitative Models (Methods)

These models usually express the numerical values of all impacts, producing a total score for all the proposed variants of the project V_i for $i = 1, 2, \ldots, m$. The BATTELLE method (Battelle Columbus Laboratories 1979) can be considered as the classic method for quantitative, HARD determinate type of analysis. There are 78 selected indicators P_j, i.e. $j = 1, 2, \ldots, 78$, divided into four categories, and an allocation is made for assessment (planning, decision-making) in the field of water resources. This method was first tested in the project of the Bear River Basin water system, extending into the states of Utah, Idaho and Wyoming. The authors identified the method as an environmental evaluation system (EES), and therefore the abbreviation EES has become somewhat synonymous with all quantitative EIA models of the HARD type.

Other important methods in this group consist of various modified procedures of discriminant analysis and the TIEQ method (total indicator of environment quality), and the AECOTEM method (aesthetic, economic and environmental impact assessment of trans-European north-south motorway).

Simulation Models

These models are recommended for predicting simple phenomena, for example transport of pollutants or requirements for natural resources. They are not recommended for initially identifying the problem; their application increases during processing of individual variants of the project.

The literature, however, indicates the possibility of applying modified approaches under the name of *Adaptive Environmental Assessment*, using simulation technology.

Models of Ecological Effectiveness

These are based on the philosophy of social ecological forecasting of the potential impact of a structure on its surroundings. Due to their nature, they seem particularly suitable for optimization of productive investments, i.e., constructions and technological processes.

Many authors have attempted to classify the various methods, applying comparison of the results and their evaluation. Summaries are presented primarily in books such as The E7 Network of Expertise for the Global Environment (1997), Walker and Johnston (1999).

Application of professional experience does not represent a strictly formal method, but many experts use knowledge and experience gained in the past in dealing with future projects.

Application of a method must always be critically assessed in terms of its suitability for a particular purpose and in terms of specific features of the given region.

1.1.4.2 Impact and Manners (Methods and Process) of Its Assessment

Strict attention is paid to the technical way of environmental impact assessment (assessing). The aim is to understand "how" and "when" to use a suitable methodology as a tool for identifying impacts and their consequences. There are three working steps (stages) of the EIA process (Říha 2001):

- detailed *identification and definition* of the impact, allowing an understanding of the nature and potential causes of its occurrence;
- *analysis* of the impact associated with the collection of the basic data, the important characteristics of a variety of analytical procedures to determine the nature, size, scope and effect of impacts;
- determining the *significance* or *acceptability* of the impact, including the necessity of its mitigation.

The process of environmental impact assessment is generally carried out in the scoping stage. In connection with this issue, the next section discusses the methods of impact prediction.

Prediction of Potential Impact on the Environment

Prediction of change (*impact*) assumes that there is a relationship between the proposed activity and the environment. These relationships can be described as a *chain of causes and effects*, which can be distinguished (Říha 2001) as follows:

- cause (direct effect associated with the activity, e.g., emissions, contaminated runoff);
- effects of first order/primary effects (mostly physical and chemical changes in the environment);
- effects of higher order/secondary effects (mostly impacts on recipients, e.g., human, flora, fauna, intangible assets, artefacts).

Potential (probable) impacts have to be assessed *in terms of their fundamental nature*, i.e., whether the impacts are:

- positive (useful), or negative (adverse);
- direct or indirect, or potentially related to that effect (primary, secondary, tertiary);
- in real time or delayed time;
- short term or long term;
- reversible or irreversible;
- local or strategic;
- physical, chemical, biological, microbiological, biochemical.

A formalized approach to assessing such greatly differentiated impacts is thus required, involving *methodological simplifying*. All the effects of implemented projects on the environment can be divided into engineering and biophysical, economic, social and cultural, health, political and aesthetic. In terms of the modern and wider conception of environmental assessment, the majority of impacts can be identified as environmental impacts. Boundaries between evaluations of aspects (categories of criteria) tend to overlap. However, the expected change can be compared with the current state, desired or standard (normative), all in terms of positive or negative (risky).

Říha (2001) distinguishes the system, multi-criteria and prognosis evaluation in the EIA process. Emphasis on prognostic considerations must not be allowed to override the fact that the quantification of indicators and criteria must *distinguish the level of descriptive evaluation from the normative evaluation*. These levels may not be confused or assimilated; but it is evident that the normative evaluation in fact presupposes the existence of available descriptive data on the grounds of what must be prevented. It follows that something that has previously not been sufficiently described from every point of view cannot be seriously or reliably evaluated.

Descriptive assessment consists of data that are measured, or in some way found, calculated, expected, estimated in an expert manner, potentially possible.

Normative assessment is based on the maximum (or if applicable minimum) limit values of international and national standards, sanitary regulations, commonly known toxicity degrees of hazardous substances, prohibitions, or acceptable risks.

Difficulties affecting compliance with the conditions which meet the assumption of ***the full process of prediction*** and assessment of impacts lead to efforts to simplify the process, while still strictly respecting the systems approach. Generally, ***an incomplete process of prediction*** is one for which two acceptable ways are proposed, i.e., both the application of approximate methods and shortening of the prediction process.

Uncertainty constitutes an undesirable component of any prediction. In terms of the EIA process, this means information about changes in the environment as a result of a proposed action. *It is axiomatically true that in each case the processed prediction will be uncertain*. Level of uncertainty depends on the input data and the applied methodology. In the literature, we find that the impact of a prediction is the "Achilles heel" of the EIA process. Generally, we can predict the occurrence of the impact, but it cannot be precisely quantified.

Differences in natural changes affect the selection of the parameters and strongly affect the accuracy of impact prediction. Care should always be taken therefore to distinguish natural changes from anthropic/human impacts.

In general, it can be concluded that uncertainty arises in the process of:

- numerical modeling and prediction of assessed impact (change in value or magnitude of parameters); and
- assessment of the significance of the effect under consideration.

Special studies related to the methodology of EIA mention some additional problems connected with the proper course of the evaluation process. The important features of partial prediction include the required *quality of information, the selected prediction method and credibility of formulated significance*.

Part of the prediction is numerical evaluation of the effect. The way depends on the nature of the effect and the recipient. ***Objective assessment*** is based on the use of objective technical and economic units, e.g., SI. ***Subjective assessment*** requires special attention and sensitive work with verbal and numerical scales. Subjective assessment generally distinguishes three ways, i.e., assessment of (Říha 1995):

- indicators (or indicator values);
- indices of value;
- direct application of interval or proportional scale.

The first method consists of only a *very approximate method*, where an indicator may represent by its value a description of the analyzed problem.

The second method expresses the opportunity to state the magnitude or quality of parameters using the *index function* of several variables, i.e. $P = f(x, y, z, \ldots)$. A team of experts can formulate and define an entirely new index function for a particular requirement. The third method uses subjective assessment of the full features of interval and ratio scales.

Characterizing the source of impact is an important indication of quantitative, qualitative, spatial (territorial) and temporal factors.

Formalized workflow involves ensuring that detection of an impact is done using a single method, and prognosis of induced changes is carried out on a scientific basis.

Predictive Methods in EIA

***Prediction** is a crucial activity in EIA*. It provides information on which the final decision about the adoption of a project variant will be based, including prevention, mitigation, compensation and restrictive measures.

***The aim of prediction** in EIA is to describe the likely changes* that will result from the proposed action or implementation of the project. This presupposes that it is possible to:

- describe the current state of the activity in the territory carried out on the basis of local investigation, or monitoring, if applicable
- perform prediction of parameter values using predictive methods,
- evaluate current and future (changed) conditions using recognized evaluation methods.

The predictive method for a particular problem is chosen based on the amount and the uncertainty of input information. Availability of input information constitutes a key factor for the choice of the method, for which the analyst is responsible. The standard, accepted methods for prediction are the following (Říha 2001):

- expert assessment both implicitly and explicitly by using expert model systems, comparisons,
- terrain experiment,
- mathematical modeling,
- visual modeling (drawing, photography, film, 3-dimensional model),
- physical modeling (e.g., noise, air pollutants, and soil profile).

Predictive methods are distinguished as *formal* and *informal*, as shown in Fig. 1.2.

Individual formal workflows produce a large category of technical-methodical prediction methods (Říha 1987).

Exact methods are performed in practice by developing a specific technical and economic plan in different variations, measurement of the data from the map, calculations and so on, mostly in technical and economic objective units.

Statistical methods are based on the collected data relevant to the specific problem and study area. They use the scientific foundations of forecasting, theory of growth curves, and assessment of the likelihood of developments.

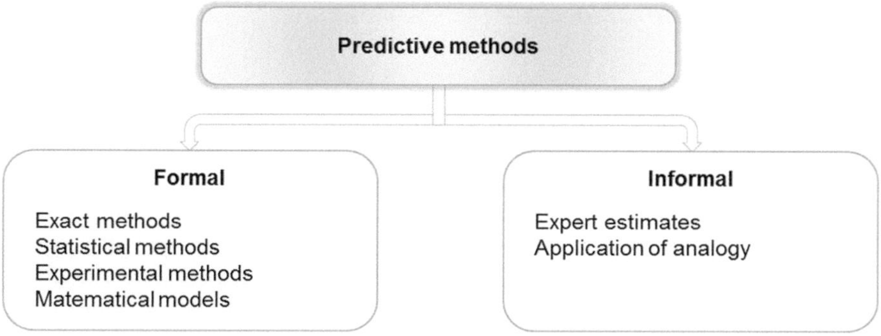

Fig. 1.2 Differentiation of predictive methods (after Říha 1987)

Experimental methods and mathematical models differ in that mathematical models explicitly formulate causes and effects, whereas in experimental methods that relationship may not be known. In both cases it is necessary to schematize (simplify) the system, resulting in a definite relation of cause → effect.

Experimental methods can be distinguished in three categories, i.e.:

- *illustrative* or *physical models*, showing the affected environment;
- *terrain (ex situ) experiments*, designed to study real changes in the study area;
- *laboratory experiments*, simulating biological and biochemical processes, but often isolated from the whole system.

In ***mathematical models***, the relationship between cause and effect is represented by one or more mathematical equations. Mathematical models can be:

- *Empirical*—based on experiments or repeated measurements, using knowledge of statistical analysis, showing the relationship between cause and effect without explicitly formulated internal relations (black box model). Apart from the general model, other models are also used specifically for application in a particular location or for a particular type of environment.
- *Processed*—consists in explicitly defining the process as such, regardless of the development in time (steady, sustained state), or otherwise (dynamic evolution). The complexity (algorithm) of the model is changed from a simple one, which can be handled manually, to complex dynamic and stochastic models that require computing.
- *Combined.*

The prognostic dimension of EIA is currently considerably strengthened and facilitated by the development of GIS categories and widely available computers.

The role of science in EIA is gaining increasing attention. EIA research has to develop if this globally important decision-making tool is to fulfil its potential as a tool for environmental management and sustainable development.

1.1.5 EIA in Slovakia

EIA procedures for public and private projects that are likely to have significant effects on the environment have been in place since the adoption of the EIA Act in 1994. In 2006, a new EIA Act was approved (Fig. 1.3), and EIA procedures began to be applied to buildings under the 2006 Planning Act. The 2006 EIA Act introduced no major changes in EIA procedures, but it tightened certain procedural time limits and better delineated EIA responsibilities between the Ministry of the Environment (MoE) and the regional and district environment offices. It also harmonized the Slovak EIA legislation with three EU directives and put preconditions on the accession of Slovakia to the UNECE Convention on Access to Information, Public Participation in Decision-Making and Access to Justice in Environmental Matters (the Aarhus Convention). The adoption of the EIA Act provided a firm basis for assessing forthcoming projects to be financed by EU Structural and Cohesion Funds.

The number of EIA procedures was low (30–70 per year) until 2000, but increased to around 200 in 2001, after the scope of projects subject to EIA was extended. Greater involvement by subnational environmental bodies after 2000 was also a factor. The number of EIAs increased further during the review period, reaching nearly 900 cases in 2008 (MoE 2010). Documentation from the assessment process is available to the public in electronic form on the website of Ministry of Environment of the Slovak Republic. The complete documentation from 20 years of experience with EIA is archived in the EIA Documentation center at the Slovak environmental agency in Banská Bystrica. Figure 1.4 summarizes the number of completed assessments of the proposed activities (EIA) in Slovakia in 1994–2012.

Non-government organizations (NGOs) have criticized EIA procedures in Slovakia for insufficient consideration of assessed alternatives, short consultation periods, limited access to the reasons underlying decisions and failure to carry out EIAs (JaE 2009).

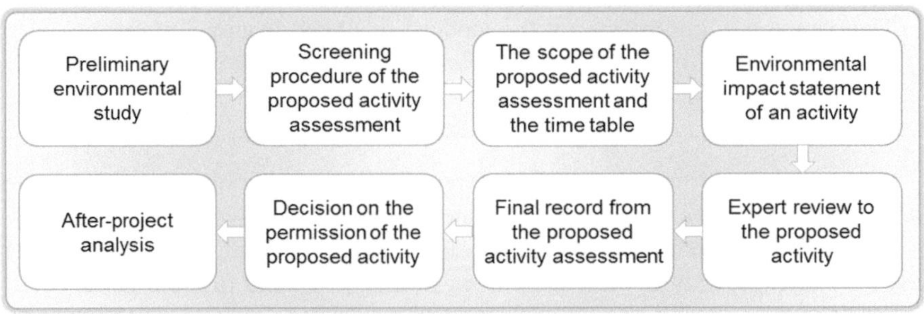

Fig. 1.3 Main steps of the EIA process in Slovakia (National Council of the Slovak Republic 2005)

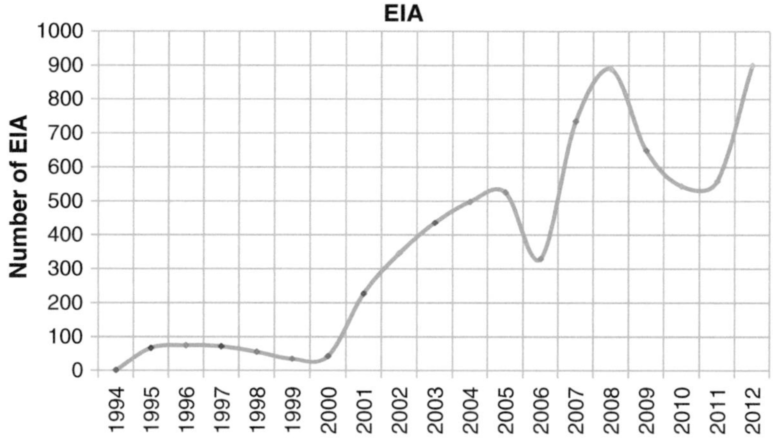

Fig. 1.4 Number of completed assessments of proposed activities (EIA) in Slovakia in 1994–2012 (MoE 2013)

In 2009, amendments to the EIA Act clarified the screening procedures for activities below existing thresholds and revised the procedures for authorization of activities subject to EIA. In 2010, as the result of an infringement procedure started by the European Commission, the Act was amended again, with a new definition of "public concerned" being adopted. The new definition includes environmental NGOs that actively participate in the EIA procedure, ad hoc public initiatives (by two or more people) and stakeholders who actively participate in the EIA procedure and can prove their interest in the activity concerned. The changes allow actors fulfilling these conditions to become parties to the relevant decision-making procedures, and to have access to related court proceedings. The new act is very much in the spirit of the Aarhus Convention. However, there is a further need to reinforce public participation, in particular at the scoping and screening stages. The screening mechanism should be simplified and clarified, for example by detailing the selection criteria and establishing thresholds, criteria or triggers. Changes are also needed to allow better access to the results of the assessment, not only "essential parts of the decisions".

The National Council of the Slovak Republic on 21 October 2011 approved Law no. 408/2011 Coll., amending and supplementing Law no. 24/2006 Coll. on environmental impact assessment. The adjustments to Law no. 24/2006 Coll. resulted mainly from the requirement of the consistent transposition of Directive 2001/42/EC on the environmental impact assessment of certain plans and programs that the Slovak Republic was enjoined to perform by order of the European Commission (formal notice). This requirement is related mainly to the subject and scope of the assessment of strategic documents and the interested public participation in the assessment of strategic documents.

The draft of the law has brought conceptual legislative changes in the area of impact assessment of the proposed activities, particularly in matters of timing of the assessment process, as well as adjustment of Annex 8 of the law, which is a list of suggested activities requiring environmental impact assessment (waste management, infrastructure, community development projects, special-purpose facilities for sport, recreation and tourism).

One significant change was the reduction of Annex. 8 of the law in Chap. 9 entitled Infrastructure, which defines the list of activities, facilities and equipment aimed at waste management subject to screening or compulsory assessment respecting the transposition of Slovakia's commitments. The availability of permits under special regulations for the proposed activities related to waste management was simplified.

The amended law facilitates the realization of the obligations of municipalities to separate and subsequently recover organic waste from households, and it will contribute to accelerating the disbursement of EU structural funds for certain types of equipment, for example presses, crushers, separating lines or composting, and it simplifies the building of waste recycling facilities, thereby strengthening the preference of recovery and recycling of waste over disposal, according to the requirements of waste management. The change consists in the fact that such facilities will not be subject to compulsory assessment without limit, but it establishes a threshold of 5000 t/year for the implementation of the screening procedure.

The amendment to Law no. 448/2012 Coll. with effect from 1 January 2013 modifies the concept of small changes in the strategic documents which are subject to screening. That change in the law on EIA removes a deficiency in Slovakian legislation, and following the notification of this modification the European Commission concluded infringement proceedings against Slovakia.

Currently it is proposed to amend the law that supplements Law no. 24/2006 Coll. on environmental impact assessment. The amendment also addresses the fact that the new EIA process should be mandatory. Subsequent approval should be conditional on respecting the outcome of the EIA, and the compliance of project documentation with the outcome of EIA should be officially verified. The Environment Ministry admits that it introduces a possible financial burden on businesses, which will be required to submit more detailed documentation. However, this procedure will otherwise streamline the procedure for businesses and the authorities, and therefore will have a positive impact.

Experience with the implementation of the Law on Environmental Impact Assessment has confirmed that the technical and urban development of Slovakia must go hand in hand from the beginning with knowledge of how much impact new activity will have in a particular area, and how potential negative impact on the environment will be minimized. EIA in Slovakia will have to maintain the ability to flexibly respond to rapidly changing economic and social conditions, as well as the open labor market in the processing of documentation, and probably closer gradual convergence with the creators of territorial-planning documentation (via SEA) (Luciak 2012).

It is expected that EIA will continue to act as an effective tool to prevent the application of investments in Slovakia that, by their degree of environmental damage, will many times outweigh their benefits (Luciak 2012). EIA law acts not as a brake, but as a professional opinion on the environmental aspect of business, which is a synergy of the industrial and environmental solutions of proposed activities. In some cases, EIA is understood only as a "mirror" to comply with legal or technical standards, which is too weak for modern environmental planning. The assessment process should take into account the public opinion, stress factors, fear of risk, and criteria increasing the quality of life. Overall, we can say that impact assessment in Slovakia still maintains professional principles, as is evidenced by the high level of documentation standards (Luciak and Nižňanský 2013).

The issue of risk in the EIA process occurs in legislative terms in the following sections of Slovakian EIA Law (National Council of the Slovak Republic 2005):

- *Annex 3*: Criteria for screening according to § 7 of the Law: 6. Environmental risks including health risks.
- *Annex 10*: Criteria for screening according to § 29 of the Law: I. The nature and the extent of the proposed activity: 8. Accident risk associated in particular with substances or technologies used, and other potential risks arising through implementation of the proposed activity.

According to *Annex 11*, among other important points, the following fall within the complex of characteristics of environmental impact and need to be processed (National Council of the Slovak Republic 2005):

- Part III. Assessment of the significance of expected environmental impacts of the activity proposed, including impacts on health.
- Part VII. Methods used in evaluating environmental impacts of the proposed activity, and the nature and sources of data obtained on the current environmental state in the area of implementation of the proposed action.

It is appropriate therefore to require determination of the degree of risk for each assessed alternative of the proposed activity, using the methods of risk analysis.

1.1.5.1 Analysis and Evaluation of the Survey in Slovakia

A total of 50 respondents filled in the questionnaire, of whom 48% had been assessing environmental impact for 10 years or more. 66% of the respondents stated that they devoted 0–25% of their work time to the process of EIA, and 94% of the respondents devoted 0–25% of their work time to the SEA process. A total of 10% of the respondents devoted 76–100% of their time to the EIA process and, in that time period, all of the respondents were dealing with the SEA process. The largest group (38%) of the respondents dealt with 0–5 sets of EIA documentation, 28% of respondents were involved in 6–10 sets of EIA documentation, 14% of respondents with 11–15 sets of EIA documentation and 18% were involved in more than 16 sets of EIA documentation in the three-year period. The majority of respondents were involved in all of the phases of the process of assessing environmental impacts. For example, 84% of respondents had a share in planning, 60% of respondents took part in surveying activities, 68% of respondents shared in the report on the evaluation and 76% of respondents were involved in the professional assessment. A total of

42% of respondents stated that they were linked to the final positions, and 10% of respondents to post-project analysis. Furthermore, some respondents also dealt with predictions of acoustic conditions.

When going through the EIA process, respondents had to deal with a whole range of environmental branches. Protected landscapes and territories and their protected zones were the most common specialization for 60% of respondents. Most of the respondents (52%) specialized in water when dealing with the EIA process—total amount, type and quality indicators of released waste waters, the place of release (recipient, public sewage, waste-water treatment plants), the origin of waste waters and the method of treatment. Biota, protection of landscapes, protection of nature, ecological stability systems, natural sciences, land use, land and ecology were the other outputs in which respondents specialized when they were involved in the EIA process. The respondents had experience in the preparation of EIA over the entire range of industrial sectors: 50% of respondents had experience in the field of infrastructure, 38% of respondents in the energy industry and 36% of respondents in the area of sport and recreation facilities and the travel industry.

Eighteen percent of the respondents considered the methodological handbooks for the EIA process as sufficient, while 34% of respondents considered them to be insufficient, and 44% of respondents thought that they are partially sufficient. Up to 82% of the respondents stated that the primary purpose of EIA is lowering the environmental impacts. Supporting decision-making process is an additional purpose of EIA for 44% of the respondents and 6% of respondents listed reducing future costs as the main purpose of EIA. Contribution of the public in the decision-making process was also given as another primary purpose for EIA.

The awareness of the public regarding the steps in the EIA process was considered by 24% of the respondents as sufficient, while 16% of the respondents felt it was insufficient, and 52% of the respondents as partially sufficient. When processing, only 18% of the respondents had a sufficient amount of data/input data and 64% of the respondents considered the amount of data as partly sufficient.

As a weakness of the present practice in the field of EIA in Slovakia, 48% of the respondents marked insufficient post-project analysis and monitoring, while 36% of respondents thought that the limited consideration of cumulative impacts was a drawback, and 36% considered subjectivity in the prediction of impacts to be a disadvantage. Other weaknesses according to the respondents were the fact that the state did not guarantee financing of the results of the EIA, but it depended instead on the goodwill of the investor, weak legal and methodological foundations, and the fact that the processing of the documentation and professional assessment was directly paid for by the applicant.

Ad hoc methods were among the most commonly used means of identification and assessment of impacts for 44% of the respondents, while for 40% of respondents these were tables and matrixes showing causes and effects, and for 28% of respondents it was the method of mapping overlay. Methods such as the evaluation of the description or professional knowledge and experience were also listed. When evaluating environmental impacts, 50% of respondents used GIS.

In the following section, there are some interesting responses and comments of respondents given to the question whether the current EIA process is satisfactory in their opinion, and what would be necessary, from the viewpoint of increasing effectiveness, to alter/change in the currently valid EIA process in the Slovak Republic.

- The actual EIA/SEA process is satisfactory; there are problems in the data foundation and partly in the ignorance of investors regarding how the EIA process can save their investment costs (plans are often made in areas which the investor bought for implementing the proposed activity, even before the basic foundations of suitability of the enactment of the plan in such location had been determined).
- An ecologist should offer an opinion regarding every activity, because assessing impacts on the environment is involved here; in practice this is not done.
- It would be necessary to eliminate political and lobbyist operations on all levels of EIA and SEA processes.
- It would be appropriate if EIA documentation were drafted solely by properly qualified personnel, as it is done, for example, with the documentation for land-planning decisions or construction activities.
- The preparation of professional methodologies that are to be used in practice is important, especially methodologies for evaluating impacts on territories of Natura 2000 and methodologies for evaluating the impacts on biotopes. Limits need setting for assessing the importance of an impact.
- It is necessary to connect the EIA process with the process used for issuing building permits or changes of the purpose of the area usage—to ensure better connection with the public and their awareness.
- The processor of the assessment and processing of the documentation are directly paid for by the applicant, so their impartiality is practically impossible.
- At present anyone has the opportunity to make a written stance in the EIA process and thus become a participant in the processes related to issuing building permits. The possibility to express oneself as a public should be limited to the people having permanent or temporary residence in the location of the planned activities.
- It is necessary to ensure feedback between the EIA process and building permits and the subsequent carrying out of the construction.
- It would be advantageous to ensure the monitoring and assessing of the fulfilment of the program of monitoring and other conditions resulting from the final position.

The last question provided the respondents with space for their proposals, notes and comments regarding the given problems. Some initiatives relating to the EIA process are therefore summarized here:

- Fewer laws and regulations, more science.
- To improve the cohesion of EIA conclusions with the construction law (from land planning through to building permit).

- To support the independence of the assessors by excluding those with contractual relations to the applicants.
- To ensure the transfer of measures from the planning to the decision-making process according to the construction law, even in cases when the result of the surveying activities is such that the assessment is not necessary.
- All documentation should be drafted professionally by properly qualified persons.
- To develop the methodology of assessing health risks resulting from noise pollution of the environment in conjunction with legislation.
- Assessment in recent times has become very formal; almost everything is taken from archives, and the contribution of the actual authors of documentation is very small, while research and truly detailed and professional assessments are absent.

These comments are part of the obtained results of the "*Assessment of the quality of the environment in the Visegrad Four* (*V4*) *Countries*" project. The objective of the project was to facilitate and promote the development of closer cooperation among V4 countries in the field of environmental assessment. The final report of the project is available at http://www.environ.agh.edu.pl/files/Assessment-of-the-Quality-of-the–Environment-in-the-V4-Countries.pdf or in AGH (2012–2013).

The study of EIA effectiveness not only revealed a number of difficulties and constraints, but has also introduced several proposals to improve the EIA process in Slovakia (Zvijáková et al. 2014).

1.2 Risk Analysis

This chapter provides a brief overview of the history, the subject and the general objectives of risk analysis. It addresses the topic of defining risk analysis from the perspective of various authors and research fields. It also gives a brief overview of standards and risk-management dimensions. A significant part is devoted to an overview of risk-analysis methods, which are still in development. The chapter concludes with a description of examples of using risk analysis in the impact assessment of activities.

1.2.1 Historical Background

"*Risk analysis*, *risk assessment*, and *risk management*" are relatively new terms; however, they are practices with lengthy histories. The first professional risk assessors were from ancient Babylon (3200 B.C.); "they were a special sect of people who served as consultants offering advice on risky, uncertain, or difficult decisions in life—such as marriage proposals or selecting building sites".

For more than a century now, risk assessment and risk management have been everyday activities of "banking, insurance, and business operations in the world's industrialized economies. Significant applications in human health and safety emerged in the early decades of this century; research on natural-hazard risks and disaster management followed" (Boroush 1998).

Risk analysis is now widely employed in many spheres of our everyday lives. It is used to inform decisions about risks linked with chemical and physical stressors (e.g. contaminants in food and water, pollution, climate change, natural disasters), biological stressors (human, plant and animal pathogens; plant and animal pests; invasive species, invasive genetic material), social and economic stressors (unemployment, financial losses, and public security, including risk of terrorism), construction and engineering (building safety, fire safety, and military applications) and business (project operations, insurance, litigation, credit). Risk analysis has become an all-pervading but often unnoticed feature of present-day society, used by government, the private and public sectors, academia as well as individuals in various communities (Arthur and Bondad-Reantaso 2012).

A number of authors have provided accounts tracing the history and modern development of risk analysis as a field of study (Covello and Mumpower 1985; Graham 1995; Paustenbach 1995; Rechard 1999; Bedford and Cooke 2001; McDaniels and Small 2004).

1.2.2 The Importance of Risk Analysis

There is a need for a general and thorough approach to justifying, explaining, demonstrating, implementing, sampling, using, and creating real skills in risk analysis in any area of the human society. Such a need is embedded in human evolution, and its importance is evident every time one needs to develop forecasts of future courses of action (Bujoreanu 2012).

All management decisions are concerned with the future, and everything about the future is, to a greater or lesser extent, uncertain (Welsh 2012).

By carrying out a risk analysis one can (Aven 2008):

- Establish a risk picture.
- Compare different alternatives and solutions in terms of risk.
- Identify factors, conditions, activities, systems, components, etc. that are important (critical) with respect to risk.
- Demonstrate the effect of various measures on risk.

This provides a basis for (Aven 2008):

- Choosing between various alternative solutions and activities while in the planning phase of a process.
- Choosing between alternative designs of a solution or a measure. What measures can be implemented to make the process less vulnerable in the sense that it can better tolerate loads and stresses?

- Drawing conclusions on whether various solutions and measures meet the stated requirements.
- Setting requirements for various solutions and measures, for example, related to the performance of the preparedness systems.
- Documenting an acceptable safety and risk level.

The importance of risk analysis is stressed by Al-Bahar and Crandall (1990), by describing it as "the vital link between systematic identification and rational management of significant (risks)". They go on to define it as a "process which incorporates uncertainty in a quantitative manner, using probability theory to evaluate the potential impact of risk". Other authors, namely Perry and Hayes (1985), agree that the objective of risk analysis is to determine the combined effect of the identified sources of risk and opportunity and highlight those that make a significant contribution to that total.

The importance of risk analysis in the evaluation of environmental engineering has been pointed out by many studies.

1.2.3 Definition of Risk Analysis

Risk analysis professionals disagree as to how key RA terms should be defined (Schierow 2004). Numerous sources and researchers have produced diverse definitions of risk analysis (see Table 1.5).

This range of definitions suggests that some authors consider risk analysis in a wider sense, i.e. their definitions comprise risk evaluation, risk management and other processes, whereas other authors see risk analysis as just one part of the whole risk management process. The above summary of the various approaches to risk analysis allows the conclusion "that risk analysis as a constituent part of risk management involves systematic application of information for risk evaluation and assessment, as well as for the selection of risk-reduction methods" (Startienė and Remeikienė 2007).

In can also be defined as "*an objective, systematic, standardized and defensible method of assessing the likelihood of negative consequences occurring due to a proposed action or activity and the likely magnitude of those consequences, or, simply put, it is science-based decision-making*" (Arthur 2008).

Strutt (1993) provides the fullest definition of *risk analysis*, setting out a procedural framework with seven stages as follows:

1. Systematic assessment (question every part of the system).
2. Identification of risks (local and global scales).
3. Assessment of risks (probabilities and consequences).
4. Establishment of tolerable levels of risk.
5. Evaluation of risks. Are the risks acceptable? Can they be reduced and at what cost?

Table 1.5 Definitions of risk analysis

References	Definitions of risk analysis
Frosdick (1997)	"Risk analysis will be deemed to be the sum of the processes of risk identification, estimation and evaluation"
Boroush (1998)	"Risk analysis is being used to evaluate and manage the potential of unwanted circumstances in a large array of areas: industrial explosions; machine part and other mechanical and process failures; workplace injuries; injury or death from diseases, natural causes, lifestyles, and voluntarily pursued activities; the impacts of economic development on ecosystems; and financial market transactions—among others"
Stoneburner et al. (2002)	"Risk analysis is synonymous with risk assessment"
ASIS (2003)	"A detailed examination including risk assessment, risk evaluation, and risk management alternatives, performed to understand the nature of unwanted, negative consequences to human life, health, property, or the environment; an analytical process to provide information regarding undesirable events; the process of quantification of the probabilities and expected consequences for identified risks"
Zeleňáková (2009)	"Risk analysis is identification of risk for human and environment, determination of probability and consequences of risk"
McVicar (2004)	"A structured approach used to identify and evaluate the likelihood and degree of risk associated with a known hazard. It leads to the implementation of practical management action designed to achieve a desired result regarding protection from the hazard. Actions taken should be proportionate to the level of the risk. This provides a rational and defendable position for any measures taken to allow meaningful use of resources and for the focus to be on the most important areas that can be controlled. Risk management requires that all possible major hazards to the matter of concern should be identified"
SRA (2004)	"A detailed examination including risk assessment, risk evaluation and risk management alternatives, performed to understand the nature of unwanted, negative consequences to human life, health, property or the environment; an analytical process to provide information regarding undesirable events; the process of quantification of the probabilities and expected consequences for identified risks"
CAC (2006)	"A process consisting of three components: risk assessment, risk management and risk communication"
Tichý (2006)	"It is the basic element of risk engineering and an *inevitable condition for any decision about risk*, so it is a basic process in risk management"

(continued)

Table 1.5 (continued)

References	Definitions of risk analysis
Šimák (2006)	"It is the process of determining risk; it allows risks to be separated into those which can be continuously monitored, and those that can be disregarded; it allows the assessment of return on funds spent on preventive measures; its importance grows with the size (extent) of the system under examination"
Phillips and Subasinghe (2008)	"Risk analysis is a tool for understanding where to focus management efforts to most effectively reduce the potential environmental effects of human activities"
ISO/IEC Guide 73: (2009), STN ISO 31000: (2011)	"Process to comprehend the nature of risk *(Note 1)—Risk analysis provides the basis for risk evaluation and decisions about risk treatment* *(Note 2)—Risk analysis includes risk estimation*"
Australian Government (2013)	"Risk analysis integrates the assessment, management and communication of risks posed by, or as a result of, gene technology"

6. Determination of reasonably practicable low risk level.
7. Determination of appropriate risk-reduction measures.

Notwithstanding the confusion, a clear definition is required for the purposes of this book. Therefore, for the purpose of this study, *risk analysis* will be defined as *the process of systematically and transparently collecting, analyzing and evaluating relevant scientific and non-scientific information about any chemical, biological or physical stressors associated with proposed activity in order to select the best option for managing the risk proposed or alternative activity to be implemented.*

The relationships between the terms used in risk management and analysis are shown in Fig. 1.5.

Despite the apparent similarity of *risk management* and *risk analysis* processes, it is necessary to understand the differences between these terms. The difference between risk analysis and risk assessment is identified in Fig. 1.6.

The literature on *risk analysis*, as well as standards and guidance documents, uses a variety of terms to describe similar concepts (FAO and WHO 2006; Hill 2005; Standards Australia 2004; USEPA 1998). The main risk-analysis terms used in this book are described in Chap. 2.

1.2.4 Risk-Analysis Methods

"Methods of *risk analysis* are the means to increase the possibility of identifying all possible risks and dangers in certain conditions," (Startienė and Remeikienė 2007).

The method of risk analysis should be characterized by the following properties (Mosler 2006):

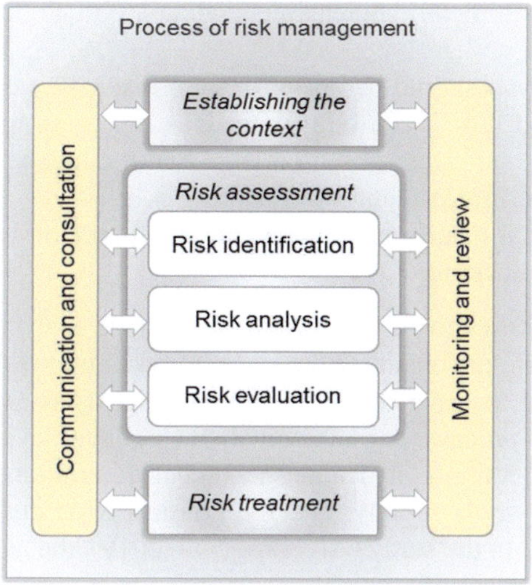

Fig. 1.5 Risk management process (adapted from STN ISO 31000: 2011)

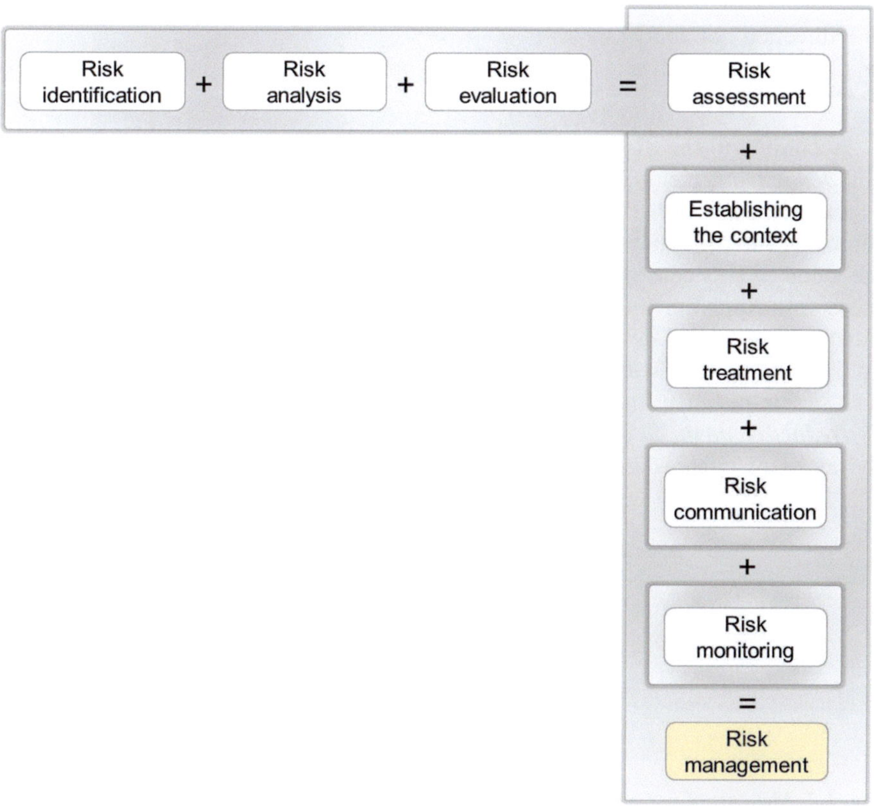

Fig. 1.6 Difference between risk analysis and risk assessment

- it should be scientifically defensible and appropriate for the system under consideration;
- the result of analysis should be in such a form that makes clearly understood the nature of risk and how it can be managed;
- it should be understandable at such a level that it could be used by experts in a repeatable and verifiable manner;
- it should provide a readily understandable interpretation of the results for subsequent decision-making.

There are many ways to describe methods of risk analysis, for example, with respect to the purpose, the result, or the system description (Johansson and Jönsson 2007). Another way to describe methods of risk analysis is to divide them into different levels, at a process, organizational and political level. In these cases, the level of the process consists of detailed technical description; the level of organization involves the whole society; and the political level contains the regulations and policy guidance. In the study (Alverbro et al. 2010), the authors decided to split the method into these specific levels.

According to Zelený (2006), there are many methods of risk analysis and risk assessment built upon different principles and practices. These methods can be grouped under three main characters, Fig. 1.7:

- "direction" of analysis
- "manner" of analysis
- "focus" of analysis.

According to **the direction of the analysis**, methods are divided into ***inductive*** and ***deductive*** (Zelený 2006; Kandráč and Skarba 2000).

Inductive methods ("*ex ante*") foresee the possible damage to devices or environment. The risk analysis points to factors that could cause impacts; help evaluate the number and consequences of impacts, and accept appropriate precautionary measures. Applying the *inductive method of analysis* means "proceeding in the direction" of logical development of the events chain, namely beginning from the collision of a risk factor with an initiating impulse through to the formation of undesirable negative impacts.

Deductive methods ("*ex post*") analyze the resulting impacts and then look for the event or circumstances that caused them. Applying *the deductive method of*

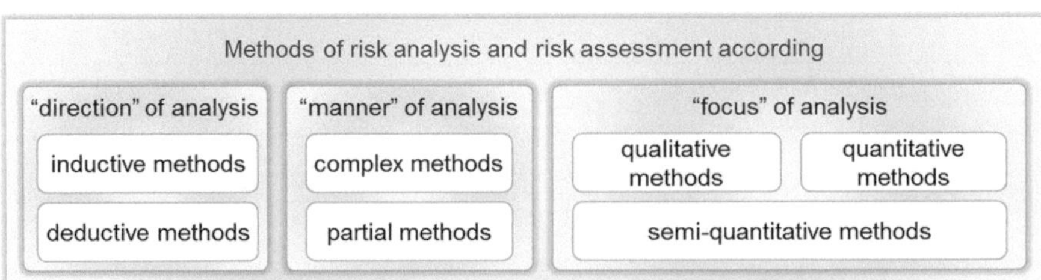

Fig. 1.7 Methods of risk analysis and risk assessment (after Zelený 2006)

analysis means the opposite approach, namely the "opposite direction" of logical development of the events chain. It begins with the undesirable negative impact, and the further procedure is intended to detect its root cause.

According to the **focus of analysis**, the methods are divided into ***qualitative*** and ***quantitative*** (Zelený 2006). *Qualitative methods* focus mainly on analysis of the logical development of the sequence of events and the whole chain of events. Quantitative methods focus on analysis and assessment of the likelihood of events and impacts. In other words, we can state that they are divided according to the manner of the description of the values with which the risk analysis works (Šimák 2006). However, there are also combined methods (***semi-quantitative***), allowing both qualitative and quantitative analysis.

A general classification of *risk-analysis* methods according to the statement of variables is depicted in Fig. 1.8.

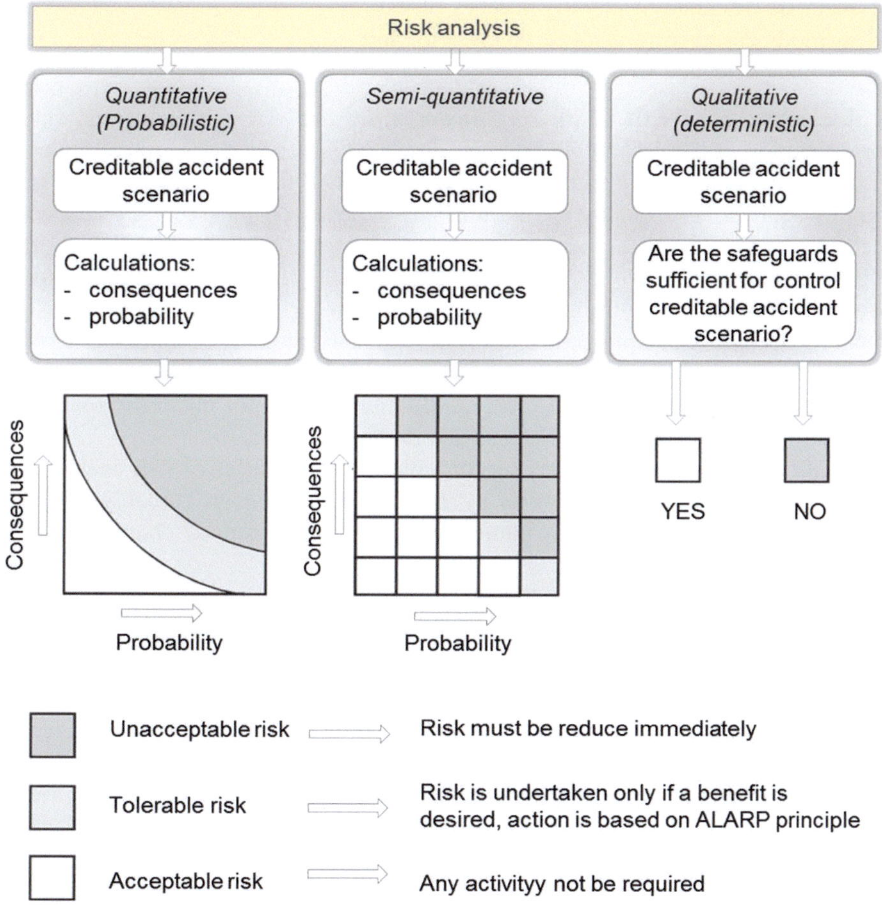

Fig. 1.8 General differentiation of risk analysis methods (adapted from Dziubiński et al. 2006)

Quantitative methods of expression of values in the process of risk analysis can be characterized as follows (Šimák 2006):

- frequency of occurrence of risk events and their possible consequences are based on mathematical terms (of the risk);
- they reflect losses (human lives and material assets—in financial terms) due to the risk phenomenon;
- greater demands for processing;
- they may be less transparent (in some cases);
- statistical analysis and simulation are used;
- special software/programs are used in which the methodology of risk analysis is incorporated.

The methods used for ***quantitative*** risk assessment can be divided into three groups:

- methods designed to identify sources of risk,
- methods designed to analyze the system, and
- methods designed to detect effects.

According to STN 01 0380: 2003, *quantitative analysis* uses numerical values for estimation of the probability and consequence (rather than descriptive scales as in qualitative and semi-quantitative analysis) obtained from the data derived from various sources. The quality of the analysis depends on the accuracy and completeness of the numerical values.

The consequences can be assessed by modeling the results of the event or set of events or extrapolation from experimental studies or from previous data. The consequences can be expressed in terms of financial, technical or human criteria or by any criteria mentioned, e.g., in STN 01 0380: 2003. In some cases, specification of the consequences at different times, in different places, for different groups and different situations requires more than one numerical value.

Qualitative methods of expression of values in the process of risk analysis are different from quantitative methods in that they are based on verbal statements, which can be transformed into numerical values. They can be characterized as follows (Šimák 2006):

- risks are expressed mainly on the basis of expert evaluation to some extent: the number of points (1–10); probability (0–1); and verbal characterization (small, medium, large);
- these methods are simpler, faster, but more subjective;
- scope of analysis is determined in most cases by a qualified estimation;
- do not allow adequate control of cost-effectiveness;
- they use mainly: Delphi method (controlled contact between experts from evaluation group and representatives of the evaluated entity); score; and brainstorming.

According to STN 01 0380: 2003, *qualitative analysis* is the narrative description or descriptive scale of the size of the potential consequences and assurance that these effects occur. Since these scales can be adapted or adjusted according to circumstances, we can use different descriptions for different risks. Qualitative analysis is used:

- as an initial monitoring activity to identify risks that require more detailed analysis;
- where the level of risk does not justify the time and effort spent on detailed analysis;
- where data are insufficient to perform quantitative analysis.

In ***semi-quantitative analysis***, the values are ranked according to the qualitative scales. The number assigned to each description may not be accurate in any relationship to the actual size of effects or credibility. Numbers can be combined according to the relationship of a number of expressions, provided that the process used to assign priorities to the selected system is responsible for the allocation of numbers and their combination. The aim is to create detailed procedures for assigning priorities other than those usually achieved in qualitative analysis, but not to propose any real value of risk, as it is in quantitative analysis (STN 01 0380: 2003).

According to the **focus of analysis**, there exist ***complex methods***, e.g., the method by which it is possible to affect all three stages in the chain of events (hazard identification, risk analysis, finding consequences), and ***partial methods*** that focus only on certain stages in the chain of events, or a maximum of two stages in the chain of events.

In practice, it is possible to meet various requirements of RA (Tichý 2006):

- **absolute analysis**; risk analysis of the investigated project is intended to determine, if possible, the exact value of risk for decision-making, with the following aims:
  - to obtain the basis for *acceptance of risk*, i.e., assess the acceptability of the proposed project,
  - to get the documents to *eliminate hazards and risks*,
- **relative analysis**—the task consists of:
  - comparison of projects in terms of portfolio risk,
  - then the decision about the project,
  - comparison of the risks within the project.

Relative risk analysis is sometimes also referred to as *preferential* or *comparative analysis*.

Tichý (2006) divides the quantification of risk, which is a segment of risk analysis, as follows:

- **analytical estimates** are based on mathematical, statistical and probabilistic analysis, and they are based on modeling the phenomena under investigation and application of the Monte Carlo method; mostly this means *absolute quantification*,
- **empirical estimates** are based on experience; this means *relative quantitation*; the empirical estimates usually use several sub-variables.

Tichý (2006) divided the tools of risk quantifying as follows:

- **Monte Carlo method**—is any simulation method based on the use of sequences of random numbers. It is useful in the evaluation of phenomena with high probability.
- **tree diagram**—can be defined as an ordered and directed graph that describes development of the events. It may also be seen as a schematic description of the process. Use requires a certain amount of imagination.
- **expert methods**—using the experience of experts in verbal or numerical estimation of the investigated problem. They are used when a decision is loaded with uncertainties.

Expert methods can be divided into two groups according to the objectives of their use:

- obtain the **verbal estimation** of hazards and risks of the project, an appraisal of potential hazard scenarios, conditions and consequences of their implementation, options for solving problems and possible options for other factors that influence decisions about risk.
- obtain a **numerical estimation** of the severity of the hazards and risks of the project, identifying highly risky scenarios or comparing two or more projects, and sometimes the possible solutions.

These methods have different applications depending on the size and complexity of the process, they produce different types of results, and they are differentially demanding on time consumption and team work. Some methods follow each other or overlap, while some are not comparable.

There are many studies about risk analysis methods. Comparison of the most common methods, techniques, methodologies, models, and tools are listed in numerous publications, including Rasche (2001), Segudovic (2006), Marhavilas et al. (2011), Reniers et al. (2005), Rouvroye and van den Bliek (2002), Raspotnig and Opdahl (2013), Altenbach (1995), Mullai (2006), and Alverbro et al. (2010).

The study by Marhavilas et al. (2011) reports the results of a classification of 404 studies focusing on techniques for risk analysis and risk assessment. They were identified through a review of 6163 papers in six scholarly journals, covering the period 2000–2009. In the study by Raspotnig and Opdahl (2013), an assessment of the results of selected techniques is evaluated according to 12 criteria.

The complexity of *risk analysis* is not helped by many authors providing diverse classifications and definitions in the various literary sources. Table 1.6 presents a classification of the risk analysis methods provided by various authors.

Summarizing the classification of methods of risk analysis, the conclusion was made that the majority of authors are predisposed to divide the methods of risk analysis into qualitative and quantitative, and that certain types of methods coincide with all classifications (for example, FTA and ETA). The state of the art of methods for risk analysis worldwide was publishing for example in Everdij and Blom (2013), FAA (2000), Canter (1998a) or Engel et al. (2006) etc.

The choice of methodology for risk analysis may involve the use of any of the four main approaches: *basic approach, informal approach, detailed analysis and combined approach* (Smejkal and Rais 2013).

It should be emphasized that there is no universally appropriate and "best" method for a particular project. Risk analysts can either use an appropriate method as it is, or they may be inspired to make adjustments. Almost daily there are new procedures, both in print publications as well as on the Internet (Tichý 2006).

1.2.5 Parts of Risk Analysis

There is a great variety of risk-analysis methods, so in order to reduce the complexity of the risk analysis, stages of risk analysis can be systemized (see Table 1.7).

This section gives the readers an overview of the various research contributions in risk-management literature. There are several different risk-management processes used in organizations today, and these are summarized in Table 1.8.

Risk analysis is usually carried out in two basic steps (Smejkal and Rais 2013):

- *Approximate risk analysis* is used for subsequent decisions about the methods (strategies) for the proper risk analysis of a particular project.
- For each project, a *detailed risk analysis* should be subsequently performed, using one of the above methods of qualitative, quantitative or semi-quantitative analysis.

Each process has its advantages and disadvantages. The starting point for choosing the optimal approach is to compare the real state of the analyzed environment with the advantages or disadvantages that come with the chosen method (Smejkal and Rais 2013).

1.2.6 The Uses of Risk Analysis and Assessment

The uses of risk identification, analysis and assessment in relation to the environment have broadened considerably in recent years. Previously, emphasis was on the

Table 1.6 Methods of risk analysis

References	Group of risk analysis methods	Types of methods
Rasche (2001), Marhavilas et al. (2011), Fera and Macchiaroli (2010), Rainer et al. (1991), Zeleňáková (2009), Flanagan and Norman (1993), Bandyopadhyay et al. (1999), Vaughan (1997), Zelený (2006), Šimák (2006)	Qualitative	What if Analysis—WFA; Fuzzy Metrics; Scenario Analysis; Failure Modes and Effects Analysis—FMEA; Failure Modes, Effects and Criticality Analysis—FMECA; Hazard and Operability Study—HAZOP; Human Error Analysis—HEA; Block Diagrams—RBD; Fault-tree analysis—FTA; Event-tree Analysis—ETA; First Order Reliability Methods—FORM; Probabilistic Risk and Safety Assessment—PRA & PSA; Monte Carlo Methods; Check-Lists; Safety Audits—SA; Task Analysis—TA; Sequentially Timed Event Plotting—STEP Technique; Questionnaires
	Quantitative	Proportional Risk-assessment—PRAT Technique; Decision Matrix Risk-assessment—DMRA technique; Risk Measures of Societal Risk; Quantitative Risk-Assessment—QRA Tool; Quantitative Assessment of Domino Scenarios—QADS; Clinical Risk and Error Analysis—CREA Method; Predictive, Epistemic Approach—PEA Method; Weighted Risk Analysis—WRA; Sensitivity Analysis; Monte Carlo Simulation; Cost—Benefit Analysis—CBA; Delphi Technique
Marhavilas et al. (2011), Fera and Macchiaroli (2010), Bandyopadhyay et al. (1999)	Combined—hybrid—semiquantitative (quantitative and qualitative approach)	Human Error Analysis Techniques—HEAT/Human Factor Event Analysis—HFEA; Fault-tree Analysis—FTA; Event-tree Analysis—ETA; Risk-based Maintenance—RBM; Attack Tree Analysis; Delphi Techniques; Value Chain Analysis

(continued)

Table 1.6 (continued)

References	Group of risk analysis methods	Types of methods
N.E.M. Business Solutions (2002), College of Engineering and Engineering Technology Northern Illinois University (2006)	Qualitative methodologies	Preliminary Hazard Analysis—PHA; Hazard and Operability Study—HAZOP; Failure Modes and Effects Analysis—FMEA; Failure Modes, Effects and Criticality Analysis—FMECA
	Techniques based on tree analysis	Fault-tree Analysis—FTA; Event-tree Analysis—ETA; Cause-Consequence Analysis—CCA; Management Oversight Risk Tree—MORT; Safety Management Organization Review Technique—SMORT
	Techniques for dynamic system	Go Method; Digraph/Fault Graph; Markov Modeling; Dynamic Event Logic Analytical Methodology—DYLAM; Dynamic Event Tree Analysis Method—DETAM
AIRMIC et al. (2002)	Upside risk	Market Survey; Prospecting; Test Marketing; Research and Development; Business Impact Analysis
	Downside risk	Treat Analysis; Fault-tree analysis—FTA; Failure Modes and Effects Analysis—FMEA
	(Both)	Dependency modeling; Strengths, Weaknesses, Opportunities, Threats Analysis—SWOT Analysis; Event-tree Analysis—ETA; Business Continuity Planning; Business, Political, Economic, Social, Technological Analysis—BPEST Analysis; Real Option Modeling; Decision Making under Conditions of Risk and Uncertainty; Statistical Inference; Measures of Central Tendency and Dispersion; Political, Economic, Social, and Technological Analysis—PEST Analysis

(continued)

Table 1.6 (continued)

References	Group of risk analysis methods	Types of methods
Frosdick (1997), Mobey and Parker (2002), Zelený (2006), Kandráč and Skarba (2000)	Intuitive technique	Brainstorming
	Inductive technique (What if?)	Preliminary Hazard Analysis—PHA; Checklists; Human Error Analysis—HEA; Hazard and Operability Study—HAZOPS; Failure Modes, Effects and Criticality Analysis—FMECA
	Deductive technique (so how?)	Fault-tree analysis—FTA; Event-tree Analysis—ETA
Tichý (2006)	Analytical estimation	Monte Carlo Method
	Empirical estimation	***One-parameter evaluation*** [***FAR Index*** (Index Fatal Accident Rate—FAR index)] ***Multi-parameter rating*** [***FN Curves*** (Fatalities/Frequncy curves—F/N curves); ***Index RPN*** (Risk Index Periodity Number—RPN index), ***Audit Risk Index*** (Audit Risk Index—AR index), ***Parameter Risk PaRs*** (Parameter Risk—PaRs)]
	Tree diagrams	Analytic Diagrams, Synthetic Diagrams
	Expert methods	Failure Modes and Effects Analysis—FMEA; Universal Matrix of Risk Analysis—UMRA, Strengths, Weaknesses, Opportunities, Threats—SWOT analysis

hazards or sources of the risk and on technical matters such as failure of machinery and industrial plants based on historical data. Consequences dealt primarily with acute effects involving human injury and death and property damage. Quantitative risk assessment (QRA) became an important tool for this task.

More recently the scope has broadened to give increasing attention to a wider range of potential consequences of hazardous events and to use QRA as a tool for land use planning and decision-making. The consequences considered now tend to include damage to flora and fauna of ecosystems (ecological risk assessment), in addition to humans and property. There has also been increasing emphasis on chronic hazards.

Risk assessment has now become an important part of overall environmental planning and management and is incorporated into a wide range of government and corporate activities to inform decision-making. The next part provides examples of some of the uses of risk assessment (Harding 1998):

Table 1.7 Parts of risk analysis

References	Stages of the risk-analysis process
White (1995)	1. "*Risk identification* Perceiving hazards; identifying failures; recognizing consequences"
	2. "*Risk estimation* Estimating risk probabilities; describing risk; quantifying risk"
	3. "*Risk evaluation* Estimating significance of risk; judging acceptability of risk; comparing risk against benefits"
Chapman (2007)	1. "*Context* Developing an intimate knowledge of the business activity under examination. Vital within the context step is the need to understand the activity objectives. The context step should also establish the "what", "when", "who" and "how" of the activity"
	2. "*Identification* Identifying the opportunities and threats to all key activities"
	3. "*Estimation* Assessing both risk and the opportunities to business in terms of their probability and their impact"
	4. "*Evaluation* Understanding the net effect of the identified threats and opportunities on an activity when aggregated"
Backlund and Hannu (2002)	1. "*Scope definition and documentation of the risk-analysis plan*"
	2. "*Hazard identification and initial consequences evaluation*, i.e., rough preliminary analysis to provide guidance as to where it is most important to begin the main analysis"
	3. "*Risk estimation*"
	4. "*Analysis verification*"
	5. "*Documentation of the risk-analysis report*"
	6. "*Analysis update* is a standard and fundamental step within risk analysis"
Rainer et al. (1991), Halliday et al. (1996)	1. "*Asset identification and analysis*"
	2. "*Threat identification and analysis*"
	3. "*Vulnerability identification and analysis*"
Suh and Han (2002)	1. "*The organizational investigation* Determining what needs to be managed; understanding the organization's mission"
	2. "*Asset identification and evaluation*"
	3. "*Treat and vulnerability assessment*"
	4. "*Annual loss-expectancy calculation*"

(continued)

Table 1.7 (continued)

References	Stages of the risk-analysis process
Mobey and Parker (2002)	1. "*Identification*, where all potential risks affecting an organization are identified"
	2. "*Estimation*, where the identified risks are assessed and their importance, likelihood, severity and impact are determined"
	3. "*Analysis and evaluation*, where the acceptability of the risk is determined and the actions that can be taken to make the risk more acceptable are evaluated"
Pačaiová et al. (2009)	1. "*Description of the system and determination of the limits*"
	2. "*Hazard identification and possible consequences*"
	3. "*Risk estimation (estimation of the probability and severity of consequence)* $R = P \times C$"
Zeleňáková (2009)	"Evaluation of the potential adverse effects that natural processes and/or human activities have on the environment (humans, plants and animals which make up ecosystems)"
USEPA (1992)	1. "*Characterization of exposure*"
	2. "*Characterization of ecological effects*"
Australian Government (2013)	1. "*Risk context*"
	2. "*Risk assessment* *Risk identification* *Risk scenario* *Risk characterisation* *Consequence assessment* *Likelihood assessment* *Risk estimate* *Risk evaluation*"
	3. "*Risk management* *Risk treatment* *Monitoring and review*"
	4. "*Risk communication*"

a. ***Setting Standards***

Quantitative risk assessment may be used to set standards for environmental management. This may involve the adverse effects of pollutants in the environment on human health. For example, an analysis may be made of the risk to human health from lead in urban air. Medical authorities will consider the risks to human health from a range of ambient air lead levels and make recommendations to government on a level which should not be exceeded in particular situations. Government in turn will consider the sources of air-born lead and may set in place standards aimed at reducing the emissions; for example, standards regarding the amount of lead in petrol. Standards may also be set in relation to risks to ecosystems, for example, the risks of algal blooms associated with release of phosphorus and nitrogen compounds into waterways.

Table 1.8 Risk-management process (after Al-Hawari et al. 2008)

References	Description of process						
	1	2	3	4	5	6	7
Boehm and Bose (1994)	Risk management mandate definition	Goal review	Risk identification	Risk analysis	Risk control planning	Risk control	Monitor risks
Kontio (1997)	Review define goals	Identify and monitor	Analysis risk	Plan risk control	Control risk		
Higuera and Haimes (1996)	Risk identification	Risk analyze	Risk plan	Risk track	Risk control	Risk communication	
Cornford (1998)	Risk identification	Risk analysis	Risk planning	Risk tracking	Risk control		
Jurison (1999)	Risk identification	Risk analysis	Risk prioritization				
Bandyopadhyay et al. (1999)	Risk identification	Risk analysis	Risk monitoring				
Bruckner et al. (2001)	Goal definition review	Risk identification	Risk analysis	Risk planning	Risk tracking	Risk tracking	
Beck et al. (2002)	Risk identification	Risk evaluation	Risk control	Risk monitoring			
Smith and Merritt (2002)	Identify risks	Analyze risks	Prioritize and map risk	Resolve risks	Monitor risks		
Sommerville (2010)	Risk identification	Risk analysis	Risk planning	Risk monitoring			
Natural Resources Commission (2013)	Risk process initiation	Risk identification	Risk analysis	Risk evaluation	Risk management and treatment	Risk reporting	Risk monitoring and review

b. ***Predicting Impacts***

Risk analysis and assessment may be used to predict impacts for EIA. For example, in New South Wales since the mid-1980s risk assessment has been a formal part of development approval procedures involving projects of a potentially hazardous nature.

c. ***Strategic Planning***

Risk assessment is also used to compare risks in order to set priorities for environmental management. Risk assessment for strategic planning purposes can take place at a number of levels. Government may compare environmental risks in order to decide on priorities for use of scarce resources available for environmental management. In considering possible risk treatment measures, governments not only take account of scientific and technical solutions for lessening risks, but, equally important, may work on improving management systems so that the safety devices in place perform to expectation and the chance of accidents from human error is minimized. Similarly, it may be important to use resources to educate the community about risks and how to minimize them. The use of risk assessment by governments for these strategic purposes is increasing rapidly at this time.

Government may also use risk analysis and assessment for cumulative EIA and hence to help in planning siting of a number of industries within a region.

Corporations likewise use risk analysis and assessment to help assess possible future financial liabilities arising from their activities and products. These may include processes which lead to pollution, land degradation or accidents (such as explosions), or products which harm human or environmental health (e.g., hazardous chemicals).

d. ***Refining Oorganizational Environmental-management Systems***

Corporations and other organisations may use risk analysis and assessment to refine their environmental-management systems and policies. Risk-analysis procedures can help identify "weak points" in the management system that require special attention in order to prevent an environmental risk event such as a chronic or acute pollution episode. Also, as new scientific information becomes available regarding the potential environmental impacts posed by an organization's activities, comparative risk assessment across these activities can help identify the most efficacious and cost-effective changes in management arrangements to prevent environmental harm.

1.3 Risk Analysis and Assessment Within Environmental Impact Assessment

1.3.1 Background to EIA and RA

The Environmental Protection Authority of Western Australia has produced a summary of the key features of environmental impact assessment:

EIA is a predictive tool that is systematically applied at the early planning and design stages of development proposals so that the government and the community can form a view about a proposal's environmental acceptability.

EIA as a predictive tool dealing in uncertainty, risk and conditions, which should be applied to control potential risks and impacts.

The environmental impacts of development can be difficult to predict. Predictions must often be made when there is still uncertainty about outcomes, be they negative or positive.

EIA is information and knowledge dependent. It is this tension about how much information and knowledge is necessary to have confidence in predictions about impacts that is at the heart of EIA (EPA 2009).

Risk assessment and EIA are similar in concept as they both deal with the prediction of the future impacts or consequences arising from proposals and uncertainties about the exact nature, probability, frequency and magnitude of impact or consequences.

Both seek to inform the decision-making process about the significance of detrimental impacts and the appropriateness of risk treatments or mitigation measures.

The European Union encourages its member states to apply risk assessment in EIA, especially to extreme events, but it provides very little specific guidance on applying this type of risk assessment or analysis.

Risk analysis in EIA has tended to be limited to human health and safety risk assessments, whereas its potential for ecosystem impact assessments is still being realized (EPA 2009).

The origins and development of EIA and risk assessment were described in Gough (1988). In that report, a relationship between EIA, risk assessment, technology assessment and social-impact assessment was proposed. A modified version is shown in Fig. 1.9.

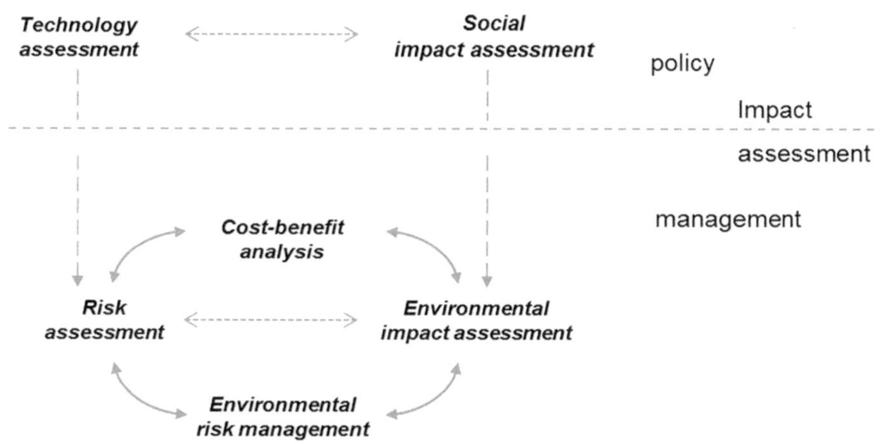

Fig. 1.9 The assessment process (adapted from Gough 1989)

Figure 1.9 shows EIA and risk assessment both contributing to environmental risk management. This contributory relationship is important because, as we have discussed, the management function goes beyond the assessment function. It takes into account the assessment and the decision and includes communication, implementation and monitoring of the selected option.

The encompassing term "impact assessment" has been used here as a generic term to describe all forms of impact assessment including technology assessment, SIA, EIA, cost-benefit analysis and risk assessment.

As shown in Fig. 1.9, impact assessment is relevant at both policy and management levels. Environmental risk management is used to describe a composite approach to the management functions of EIA and risk assessment. There are several advantages to considering EIA and risk assessment as independent components of a combined assessment-management process (Gough 1989).

EIA and *risk assessment both* involve the prediction of consequences resulting from human activities or planned interventions (Brookes 2001). They are both intended to provide information to decision-makers about adverse consequences and their significance, and as decision-supporting tools they should support decision-making on measures to reduce or eliminate adverse impacts.

Enhanced integration of risk assessment (risk analysis) in EIA may contribute significantly to moderating the risk of environmental disasters (Lexer et al. 2006). EIA has firm roots in legislation and regulatory frameworks; whereas there is no strong historical tradition of integration of risk assessment approaches into environmental policy and legislation (Brookes 2001).

Nevertheless, risk assessment and management approaches have become increasingly important for environmental policy in recent decades (Fairman et al. 1999). Examples of complete environmental risk assessments being applied in EIA have been rare on a-n European level, but certain risk analysis techniques on the other hand have often been used in EIAs, particularly techniques of human-health risk assessment, e.g. pollutant distribution and exposure paths, and safety risk assessments are relatively common for specific accident-sensitive project types involving high-risk technologies (e.g. nuclear power plants and chemicals industries).

The World Bank as well as the Asian Development Bank have issued EIA guidelines giving detailed guidance on applying risk analysis in EIA (WB 1991; ADB 1997), and they integrate risk analysis into their project preparation and implementation processes (Kjorven 1998). Lexer et al. (2006) also focused on using risk assessment in EIA. They examined the extent of extraordinary (non-routine) hazards and the ways European Union member states handle risk assessment in EIA, both within the regulatory framework and in EIA practice. This report clearly assumes that risk assessment offers many benefits as a supporting technique in EIA (Brookes 2001), and that risk-based contingency plans may even be required in many cases to achieve the substantive objectives of EIA as a tool for preventative environmental protection.

1.3.2 Similarities Between Environmental Risk Assessment and Environmental Impact Assessment

The framework of environmental risk assessment (ERA) can be combined with the basic EIA process. Many principles of the basic EIA procedure and the ERA framework coincide (Fig. 1.10, Table 1.9), and this means that integration is

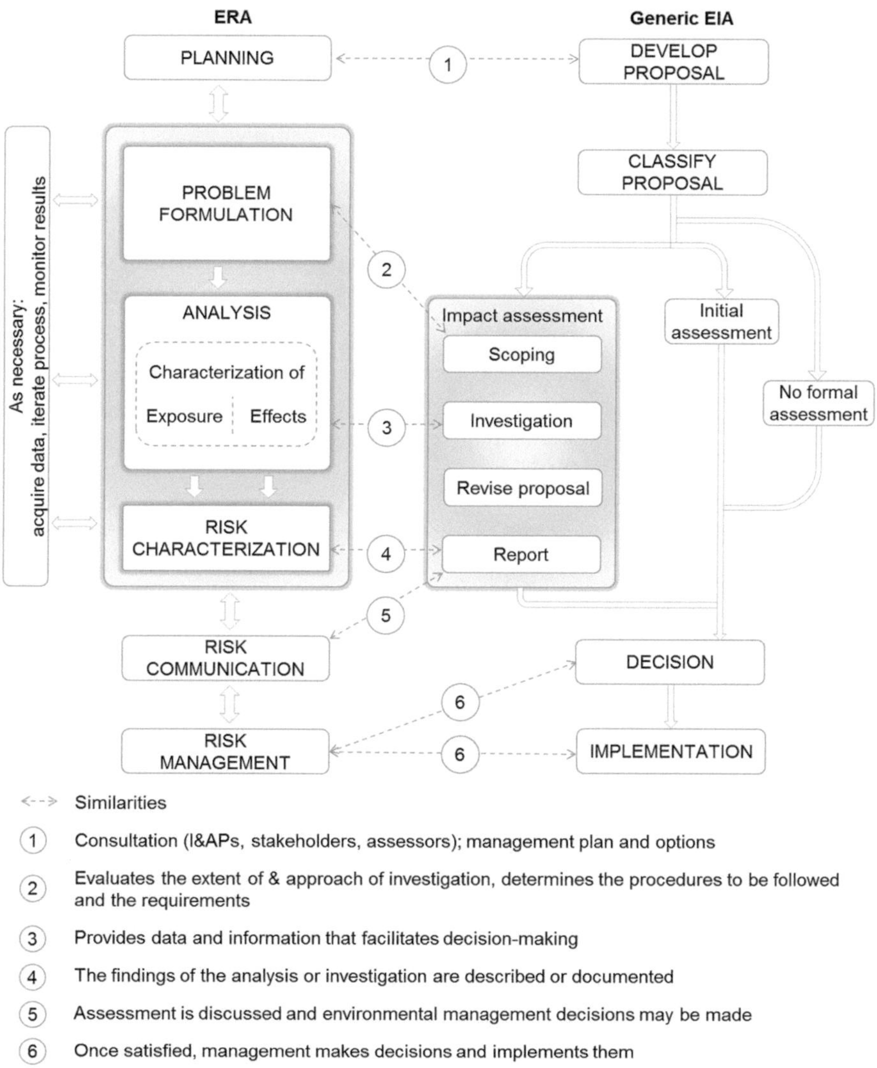

Fig. 1.10 Comparison of the ERA framework and basic EIA procedure (adapted from DEAT 2002)

Table 1.9 Similarities between the principles of EIA procedure and the ERA framework (adapted from DEAT 2002)

EIA	ERA
"Accountability for information and decisions taken"	"Risk manager is accountable"
"Open, participatory approach"	"Participatory approach from planning to risk communications"
"Consultation with interested and affected parties"	"Risk communication occurs with interested and affected parties"
"Considers alternative options"	"Alternative options are considered in remediation approaches"
"Ensures social costs of developing proposals will be outweighed by social benefits"	"Includes cost-benefit analysis"
"Opportunity for public and specialist input in decision-making"	"Risk communication between risk managers and public/I&APs in decision-making"
"Includes uncertainty"	"Includes uncertainty"

possible at all policy and regulation levels. For example, both procedures (Antunes and Santos 1999):

- "aim to balance socio-economic development objectives with environmental quality and ecological functions to promote sustainable development,"
- "assist in the development, implementation and evaluation of policies that promote sustainable development,"
- "can be applied to different levels of analysis (e.g. local, regional, continental and global scales)," and
- "are adaptive and considers problem assessment, policy priorities, formulation and implementation of policies through adequate tools, and takes into account the perspectives of the stakeholders involved."

These two instruments complement each other because EIA focuses on all the impacts identified in a specific project, whereas ERA provides a structure for approaching ecological issues above all. ERA is based on fundamental risk assessment principles, so this approach is effective for pinpointing risks right from the planning stage. Following a risk-based approach in ERA improves the scientific rigour and soundness of the process. Moreover, incorporating ecological factors into EIA enables their interaction with social and economic factors to be assessed.

Recognition of the insufficient role of health impacts in EIA has led to discussion of the need to integrate RA and EIA processes. Many authors (e.g., Grima et al. 1986; Andrews 1990; Arquiaga et al. 1994; Canter 1996, 1998b) see the fuller assessment of health impacts in EIA as a product of using "scientifically-based" risk analysis methods (Demidova and Cherp 2005).

An integrated environmental management procedure naturally promotes a holistic, interconnected approach to environmental system control using a goal-oriented process (Antunes and Santos 1999). The ERA process itself also supports this philosophy.

Practice-informed EIA procedure has been successfully applied in coastal zone and catchment area management (Argent et al. 1999). Similarly, practical ERA techniques may be widely applied within EIA procedure. For example (DEAT 2002):

- in determining acceptable risks to develop environmental standards,
- in site-specific decisions (hazard identification or land-use planning), and
- in comparative risk analysis (compare different types of risks, make alternative risk options).

Risk assessment and management may be used either as alternatives (especially applied to existing projects) or as part of EIA, because risk assessment is able to deal with specific impacts involving a higher degree of uncertainty than other, more easily-predicted impacts (DEAT 2006).

1.3.3 Existing Approaches to Integrating RA into EIA

EIA and RA share an ultimate goal, "the rational reform of policy-making" (Andrews 1990). Both assessment tools are intended to produce reasoned predictions of possible consequences of planned decisions to enable better choices to be made from among the available options. However, the tools are products of different disciplines and professions, so they have different emphasis regarding their particular substance and process. Specifically, EIA focuses on identifying impacts which could result from planned projects, whereas RA involves rigorous analysis of those impacts, i.e. calculation of the probability and severity of their effects. EIA generally emphasizes impacts on the natural environment, and while RA traditionally targeted human health issues, it has relatively recently been extended to cover ecological effects. In practice though, these distinctions do not represent hurdles to integration, but rather indicate opportunities for mutual enrichment of these two assessment systems (Demidova and Cherp 2005).

The following reasons can be proposed for integrating RA and EIA. The initial idea was that EIA could benefit from incorporating RA approaches specifically to improve the assessment of human-health issues. Then with regard to the uncertainty of impacts, RA has been promoted as a way of making impact prediction and evaluation more rigorous and scientifically justifiable. In addition to impact analysis, RA can make analysis of alternative impact mitigation strategies clearer and more effective. The benefit for impact assessors is evident, but EIA/RA integration would also provide for "greater clarity and transparency in decision-making" (Eduljee 1999) and facilitate risk management during the implementation of projects. Moreover, integration should support the establishment of RA procedure as a natural component of EIA as a widely-used decision-making support tool. RA would in turn be enhanced with public participation and consultation elements borrowed from EIA (Demidova and Cherp 2005).

Various ways of integrating risk analysis into the EIA process have been suggested in the last two decades (see e.g. AGIP KCO 2004; Demidova and Cherp 2005; Catchpole and Moreno 2012). Nevertheless, no generally-accepted model of RA/EIA integration has appeared so far. Petts (1997) saw the main problem impeding integration as the predominant view of RA as a separately-developed safety-supporting procedure which could and should be used in parallel with EIA. There are examples of RA methods being practically applied in EIAs of high-risk, high-profile projects with human-health implications requiring assessment, including nuclear-waste disposal sites (Arquiaga et al. 1994), toxic and hazardous waste incinerators (Petts and Eduljee 1994), and municipal-waste incinerators (Petts and Eduljee 1994; Harrop and Nixon 1999). The lack of a comprehensive methodological framework has resulted in “a generally *ad hoc* approach to RA in the context of EIA” (Eduljee 1999).

Shine et al. (2000) describe using risk analysis and EIA in permit systems. Where available, internationally agreed standards and norms should be followed in the design and content of assessment procedures. National legal frameworks should clearly specify that risk analysis and EIA must be carried out prior to the determination of a permit application. They should provide for the issue of regulations setting out appropriate methodologies, criteria and administrative aspects, as well as the information to be supplied by the applicant to the competent authority (Shine et al. 2000).

Relatively few countries have a comprehensive legal basis to conduct risk analysis of proposed development projects. In drawing up the necessary regulations, lawmakers should give special consideration to the need for flexibility and regular updating in line with scientific developments. The IUCN Guidelines provide a useful checklist of generic questions that should be applied in risk analysis (see IUCN 2000). Increasingly risk analysis is being seen as a key tool to assist in screening, scoping and decision-making (GESAMP 2008). In South Australia, there is now strong emphasis on a risk-based approach to assessment in water management. A risk profile of proposals is developed depending on the manner in which water is discharged (none, controlled, or uncontrolled) and the amount of feed input (natural; minor manufactured; major manufactured). This risk profile is used to determine the scope of the assessment and need for mitigation or monitoring. Risk-based approaches are also being promoted in New Zealand. It should also be recognized that the risk-analysis approach has often been used implicitly in screening and scoping procedures for many years (FAO 2009).

The following is a presentation of the different approaches of integration RA into EIA:

a. AGIP KCO (2004) presented the suggested methodological aspects of EIA to be used for overall EIA development. The methodology developed for EIA is based on the definition of the following three parameters:

- spatial impact scale;
- temporal impact scale;
- impact intensity.

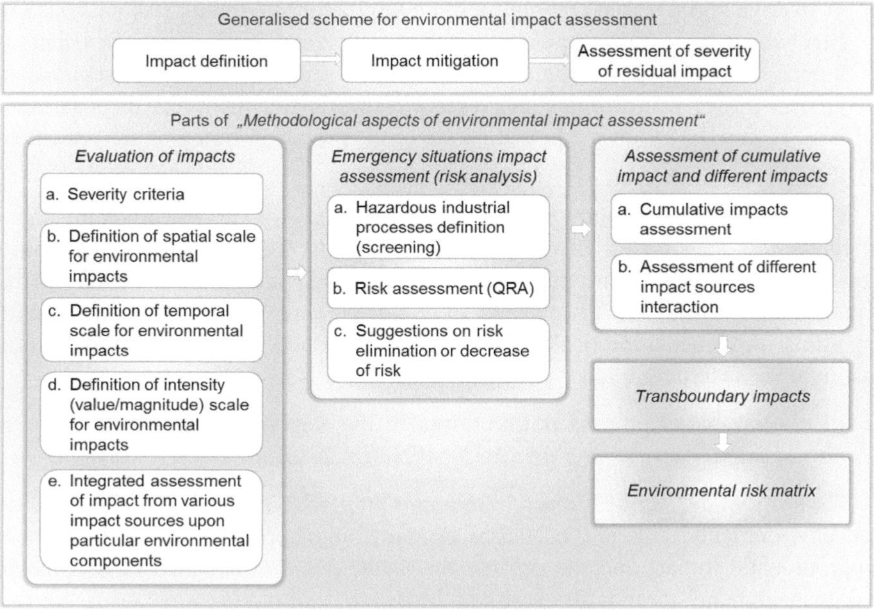

Fig. 1.11 Methodological aspects of EIA (after AGIP KCO 2004)

Each of the impact scales is scored on a 4-point gradation. The appropriate criteria have been developed for each of gradation. The generalized scheme for EIA is show in Fig. 1.11. The assessment of the potential impacts of the project upon the environment is the most important stage in the EIA process. The objective is to define what environmental changes may result from the proposed activities and assess the severity of such changes. This assessment is based on the following aspects:

- technical description of the project;
- understanding of the environmental components that may be affected;
- experience from other projects.

The procedure for determining an integrated impact assessment for environmental components for routine operation and emergency situations is presented. A mechanism for defining the impact severity is suggested. The developed methodology for EIA enables:

- the assessment of environmental impacts from different sources for both routine and non-routine (emergency) operations;
- the severity of environmental impact to be defined;
- the assessment of cumulative impact and interaction of different impact sources;
- transboundary impacts to be assessed.

The suggested methodological aspects allow making certain conclusions on the assessment of impact on the environment (AGIP KCO 2004).

b. Demidova and Cherp (2005) proposed a model for integrating risk analysis into EIA which was still a framework for assessment (albeit more consistent) of human-health impacts of high-risk, high-profile projects, including chemical and nuclear power plants, dams and reservoirs, waste treatment and disposal facilities. Figure 1.12 shows a flow diagram of the proposed "integrated" assessment process.

The findings of their study can be seen as proposals for stronger implementation of risk assessment within EIA for high-risk projects. The proposed evaluation of how risk assessment is currently incorporated into EIA may be used as an internal "quality control tool" to review and improve ongoing EIAs. A feasibility study regarding implementation of the proposed measures and the current obstacles (e.g. institutional, resource-related, political) to doing so is also recommended.

c. EPA's risk-based process is introduced in the Report: *Review of the environmental impact assessment process in Western Australia* (*March 2009*).

The section: the *Applications of risk-based approach in all stages of EIA*, define: "an environmental risk is the chance of something happening that will have an environmental impact and the process that addresses risk is called environmental risk management" (Standards Australia 2004).

Environmental risk management and EIA processes both involve consulting stakeholders and establishing the context of the assessment. The risk-based EIA in Western Australia involves melding both processes to utilize the benefits of risk analysis while satisfying statutory assessment processes as prescribed by law, and encouraging the application of contemporary environmental protection practices.

Figure 1.13 provides an outline of the risk-based EIA process divided into three broad phases corresponding to stages in the conventional EIA process for proposals—scoping, detailed impact assessment and mitigation (EPA 2009).

Within each of these mentioned phases, risk analyses and evaluations of environmental risks are performed on an increasing level of detail as a proposal proceeds through the process. For instance, at the scoping stage a preliminary risk assessment would be performed using more generic consequence tables, whereas at the detailed impact assessment stage more detailed consequence tables would be constructed at factor and/or species levels depending on risk.

The section: *Risk-based assessment terminology, definitions discussion paper prepared for the EIA Review*, provides definitions of terms to be used in risk-based EIA based on a combination of terminology used in AS/NZS 203:2006 and AS/NZS ISO 14001 Standards. A schematic representation of the relationship between various proposed terms is illustrated in Fig. 1.14.

The terminology applied in risk assessment and EIA are different but translatable as shown in Fig. 1.14. The risk based EIA approach in Western Australia applies a combination of these terminology but retains terms commonly applied in the conventional EIA process, while introducing additional risk management terms as required and as defined in HB 203:2006 "Environmental Risk Management—Principles and Process" (EPA 2009).

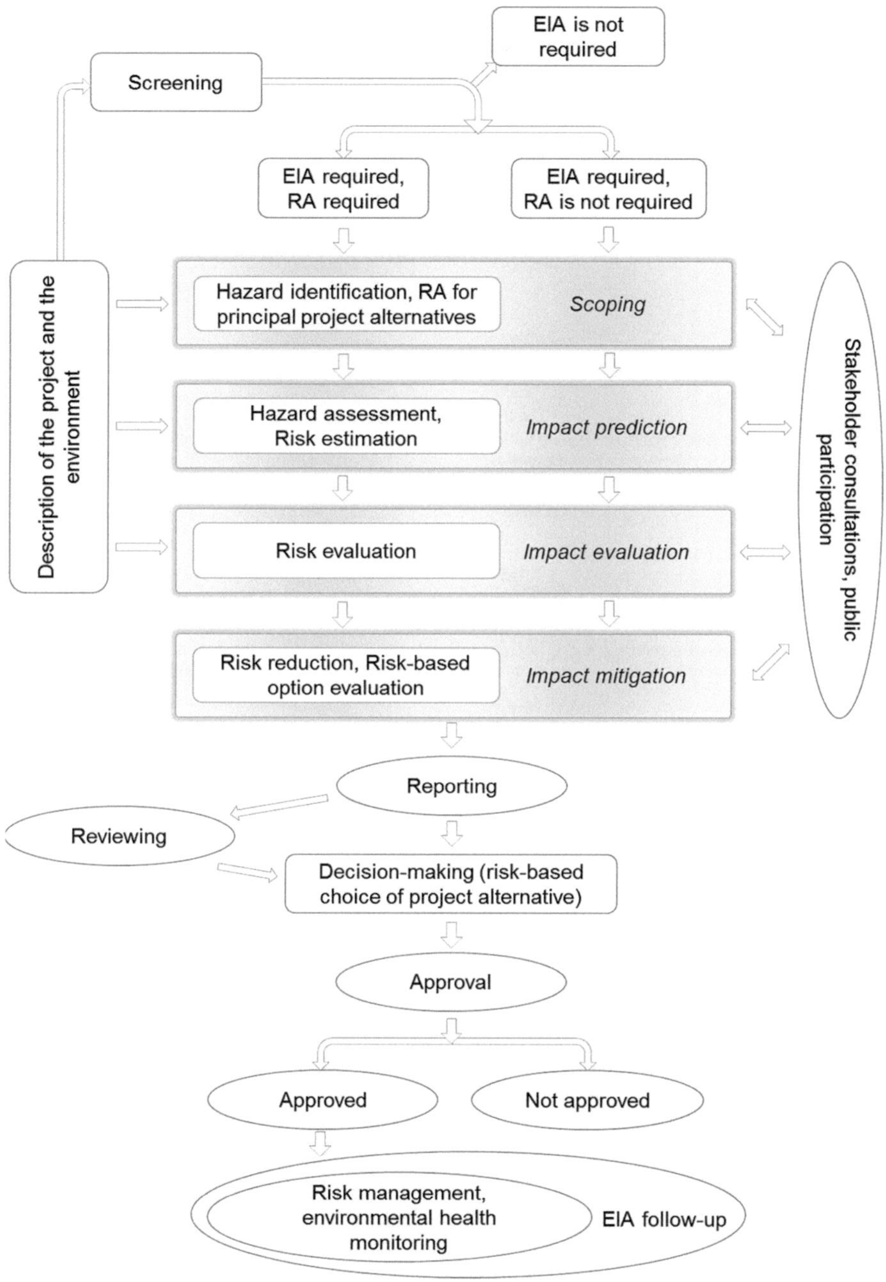

Fig. 1.12 A model for integrating risk assessment into EIA for projects with significant health implications (adapted from Demidova and Cherp 2005)

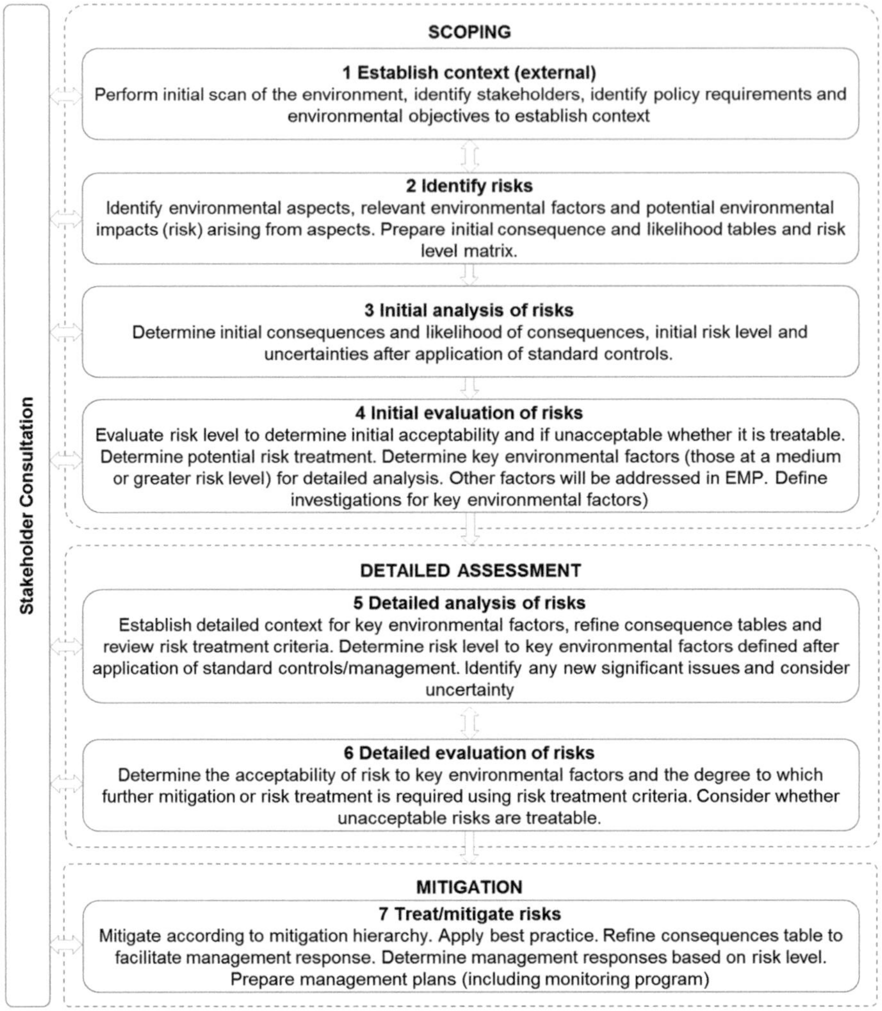

Fig. 1.13 Risk-based EIA process (after EPA 2009)

d. USEPA has stated that a risk based approach may be applied in the EIA process in Western Australia. It is assumed that this process will apply during the scoping phases of the EIA and include the consideration of potential impacts of activities on humans (Department of Health 2010).

To integrate HIA processes within EIA (Fig. 1.15), it is appropriate that the criteria used within the EIA risk based approach are consistent with the terminology and understandings used within the health sector and that they correspond with those used by the USEPA.

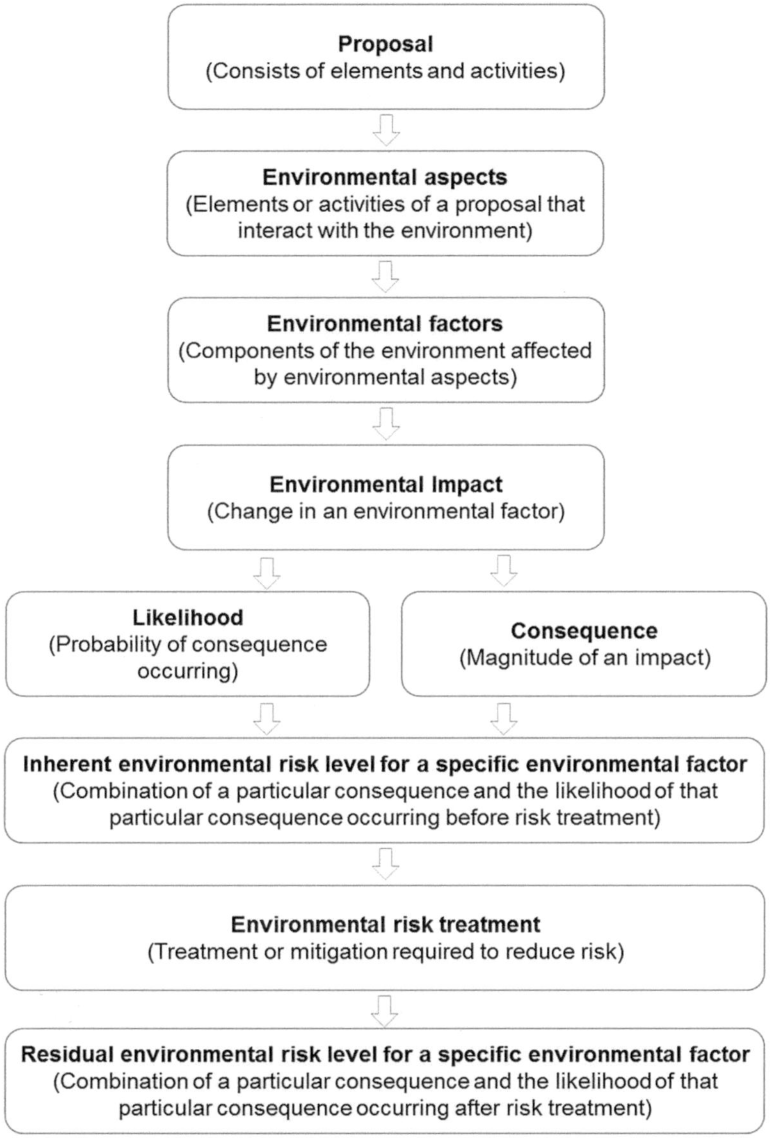

Fig. 1.14 Relationship between risk-based EIA terms (after EPA 2009)

Guidance document from Department of Health (2010) provides a framework for the health risk assessment component of the scoping phase within the processes of the Environmental and Health Impact Assessment.

e. Assessing the potential environmental impact of alien plants and plant pests is difficult. New protocols have been developed in the framework of the EC project PRATIQUE to provide guidance on environmental impact assessment in the

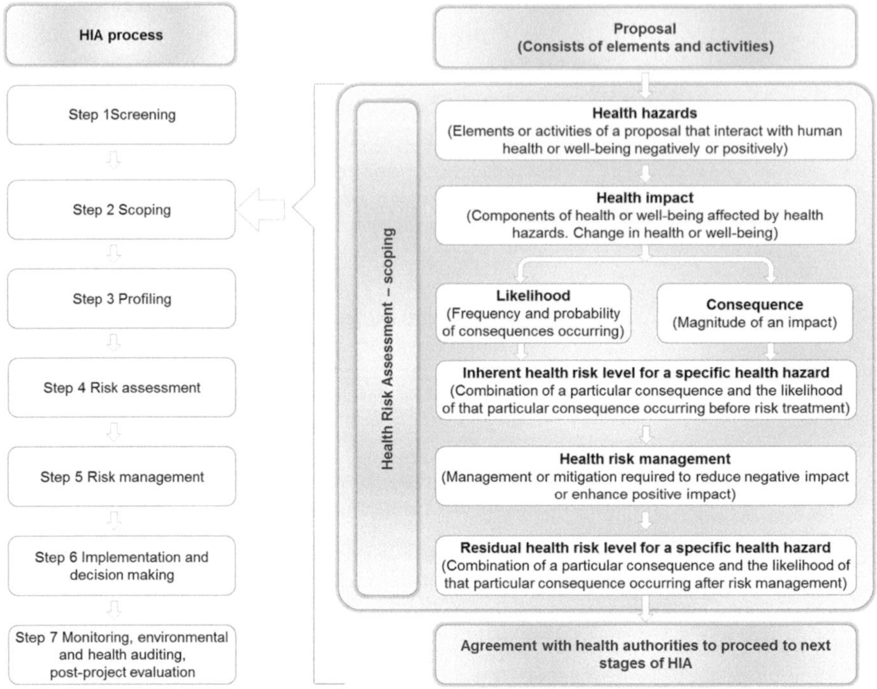

Fig. 1.15 Flow chart of the HRA process. The HRA outlined in this document is to be applied at the scoping stage of the HIA process (adopted from Department of Health 2010)

EPPO pest-risk analysis (PRA) decision support scheme and enhance consistency between risk assessors and risk ratings for different pests. A set of questions with rating guidance and examples is developed, and individual scores are summarized into final scores, using a hierarchy of risk matrices, to assess current and potential environmental impacts. Two separate protocols are available: for alien plants and for other pests. These protocols could also be used to assess environmental impact in other PRA schemes or to prioritize species for management decisions (Kenis et al. 2012).

f. The following flow diagram in Fig. 1.16 presents the general methodology adopted for various environmental studies (DESEIN 2014): Environmental Impact Assessment (EIA), Environmental Management Plan (EMP), Disaster Management Plan (DMP) Terms of Reference (ToR), Risk Analysis (RA).

g. The purpose of DEAT (2006) is to fit an additional methodology of environmental risk management to the existing series of documents presenting integrated environmental management information, which can be used to identify, evaluate and control the various types of risks which are often not covered by the traditional environmental management tools such as EIA.

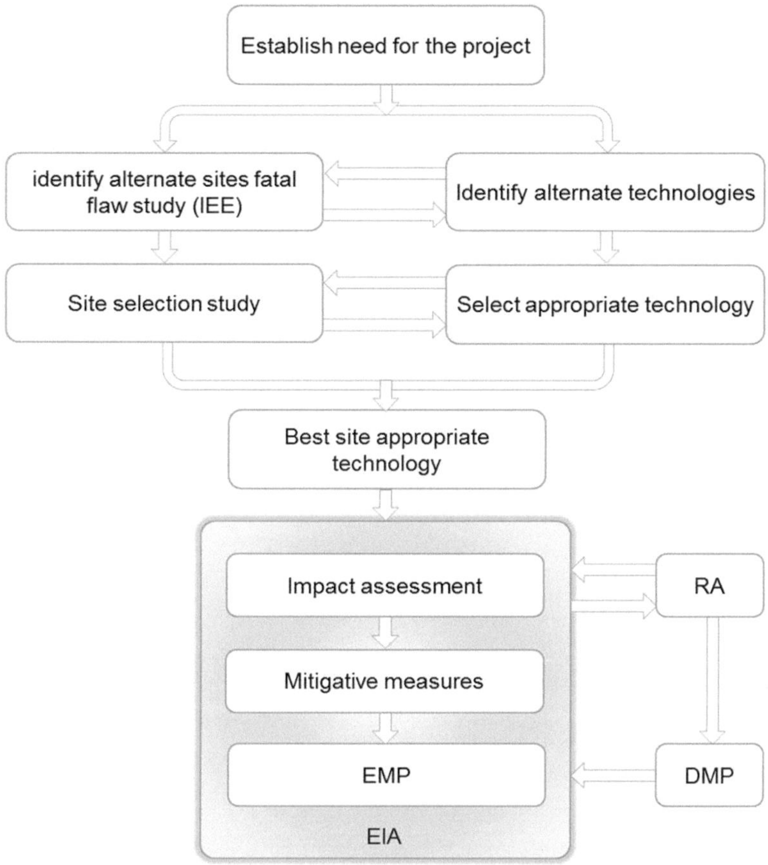

Fig. 1.16 Flow diagram presents the general methodology adopted for various environmental studies (after DESEIN 2014)

Figure 1.17 shows an overview of the risk management model in flowchart form, setting out the various terms used by the different disciplines, and giving some idea of the range of the various methods and procedures involved in the process.

In most countries, EIA is the main existing and legally required assessment tool, and many of the elements of risk analysis are already included in the EIA process, although associated with somewhat different terminology (Phillips and Subasinghe 2008).

Risk analysis should therefore be part of EIA, rather than considered as a separate or even parallel process. It is also emphasized that the risk analysis process

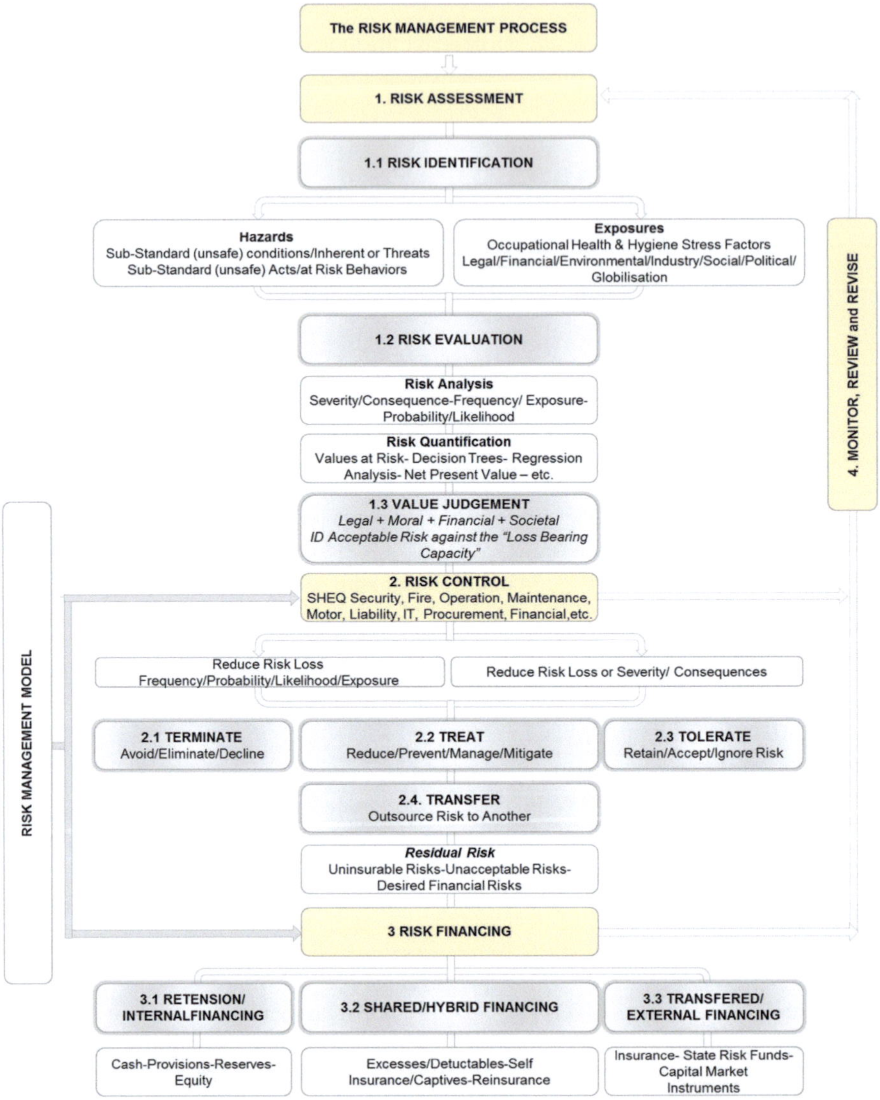

Fig. 1.17 The risk management process and risk management model flow chart (after DEAT 2006)

(as for EIA) needs to be related to management and decision makers. The analysis is of limited practical use if there is no management framework suitable for addressing the most important environmental risks associated with water management development.

1.4 Summary and Implications

Opinions on the potential uses of risk analysis and environmental impact assessment are different. Although many experts, researchers, practitioners and policy-makers agree that risk analysis is a valuable tool to inform decisions, they disagree about the extent to which risk estimates may be biased and should be allowed to influence public policies to protect health and the environment and to be used within the EIA process.

Many experts, practitioners, some academics argue that risk analysis is objective and reflects sound science. They argue it should first be used to target goals to address the worst risks to health and the environment, to achieve risk reduction in more cost-effective and flexible ways that minimize overall economic impacts, and to ensure that risk reduction achieved by regulations is worth the cost.

Other experts, some academics, and many environmentalists argue that excessive reliance on risk analysis to evaluate problems and solutions related to human health and the environment, especially quantitative risk analysis, ignores other important facets of policy decisions, such as environmental impacts, timeliness, fairness, effects on democratic rights and liberties, practicality, morality, reversibility of effects, regulatory stability, flexibility, or aesthetic values. Critics charge that quantitative methods cannot assess very long-term or newly discovered threats. They also believe that quantitative cost-benefit analyses (which are derived in part from risk analyses) undervalue environmental and health benefits, exaggerate costs, and focus on relatively widespread but individually small costs and risks, rather than on much larger costs and risks to smaller, and often more vulnerable, groups. In addition, critics charge, risk analysis typically evaluates data for well-studied and relatively well-understood hazards, ignoring emerging concerns about hazards that are poorly understood.

The quality (and value) of any risk analysis depends on the adequacy of the data and the validity of the method. For environmental hazards and most health and ecological effects, data are limited, methods are controversial, and consequently, quantitative risk estimates are very imprecise and highly uncertain, offering little guidance to policy-makers (Schierow 2004).

EIA offers an appropriate legal and procedural framework for the integration of risk assessment (Greiving 2005). The question is then why risk assessment has found only limited application in European EIA practice. One reason may be a widespread belief among the EIA community that risk assessment is complicated, difficult to communicate, and usually costly.

Some key findings from desk/empirical research from Lexer et al. (2006) are:

- No specific guidance exists on how to apply risk analysis in EIA.
- Literature on the linkage between risk analyses within EIA is rare, and few empirical evaluations of EIA have dealt with the issue up to now.

- Most practical applications of risk assessment in EIA are human-health risk assessments and technological safety-risk assessments.
- Risk analysis in EIA is often hazard-focused and based on a risk management approach.
- Diverse barriers to more coverage and deeper integration of risk analysis in EIA are identified, including:
  - lack of specific guidance, methodologies, expertise, and training;
  - missing legal requirements;
  - missing definition of the concept of risk in the context of EIA;
  - lack of adequate methods;
  - difficulties in integrating outcomes of risk assessment in decision-making processes, in particular with regard to evaluating acceptability of risk;
  - difficulties in communicating risk issues and handling them in public participation;
  - fears about increases in duration and cost of procedures;
  - lack of awareness for significance or probabilistic nature of many hazards.

However, a risk-based approach has the potential for a number of advantages including:

- greater transparency in decision-making processes;
- support informed, consistent and defensible decision-making;
- consistent with the precautionary principle;
- more systematic approach to evaluating the magnitude of environmental impacts;
- prioritizes the environmental impacts of concern, the application of management and controls and the focus of audit programs;
- improves environmental accountability of proponents;
- provides an effective basis for the engagement of key stakeholders to influence environmental outcomes;
- provides a sound basis for the development of targeted research and development programs.

The wider use of risk-based approaches is recognized as potentially helpful to define more precisely the environmental risks and enabling focus in key issues in environmental management and monitoring (GESAMP 2008).

The aim of this book is to develop a general methodology for the analysis and evaluation of environmental impacts of actions in water management (specifically FPS in Kružlov village) on the environment using the risk analysis method. The application of the developing methodology for the process of EIA will proceed as assumptions for further improvements and respectively more effective implementation and performance. The main objective of the book is to hasten and improve the evolution of the EIA process.

References

ADB (Asian Development Bank) (1997) Risk and uncertainty in EIA. In: Lohani B et al (ed) EIA for developing countries in Asia. Asian Development Bank, Manila, pp 5-1–5-45

Adelle C, Weiland S (2012) Policy assessment: the state of the art. Impact Assess Proj Apprais 30(1):25–33

AGIP KCO (2004) Regulatory basis of environmental impact assessment current environmental status. Methodological aspects of environmental and socio-economic impact assessment. Methodological aspects of environmental and socio-economic impact assessment. http://www.unece.org/fileadmin/DAM/env/eia/documents/CentralAsiaGuidelines/Annex%204%20-%20 English.pdfAGH (2012–2013) Visegrad Standard Grant No: 21210018. Assessment of the quality of the environment in the V4 Countries. http://www.environ.agh.edu.pl/

AIRMIC, ALARM, IRM (2002) A risk management standard. The association of Insurance and Risk Management, The association of insurance and risk managers, The national forum for risk management in the public sector, United Kingdom

Al-Agha MR, Mortaja RS (2005) Desalination in the Gaza Strip: drinking water supply and environmental impact. Desalination 173(2):157–171

Al-Bahar JF, Crandall KC (1990) Systematic risk management approach for construction projects. J Constr Eng Manage 116(3):533–546

Al-Hawari SF, Thabtah L, Karadsheh, Wa'el Musa Hadi (2008) A risk management model for project execution. In: Proceedings of the 9th International Business Information Management Association Conference (IBIMA), Conference on Information Management in Modern Organizations: Trends & Challenges, pp 887–893, 4–6 January, Marrakech, Morocco

Altenbach TJ (1995) A comparison of risk assessment techniques from qualitative to quantitative. Lawrence Livermore National Laboratory, Honolulu, pp 1–25

Alverbro K, Nevhage B, Erdeniz R (2010) Methods for risk analysis. TRITA-INFRA-FMS, Stockholm

Andrews RNL (1990) Environmental impact assessment and risk assessment: learning from each other. In: Wathern P (ed) Environmental impact assessment: theory and practice. Unwin Hyman, London, pp 85–97

Antunes P, Santos R (1999) Integrated environmental management of the oceans. Ecol Econ 31(2):215–226

Argent RM, Grayson RB, Ewing SA (1999) Integrated models for environmental management: issues of process and design. Environ Int 25(6–7):693–699

Arquiaga MC, Canter LW, Nelson DI (1994) Integration of health impact considerations in environmental impact studies. Impact Assess 12(2):175–193

Arthur JR (2008) General principles of the risk analysis process and its application to aquaculture In: Bondad-Reantaso MG, Arthur JR, Subasinghe RP (ed) Understanding and applying risk analysis in aquaculture. FAO Fisheries and Aquaculture Technical Paper. No. 519. FAO, Rome, p 304

Arthur JR, Bondad-Reantaso MG (2012) Introductory training course on risk analysis for movements of live aquatic animals. Samoa, FAO SAP, p 167

Ashby E (1976) Background to environmental impact assessment. In: O'Riordan T, Hey R (eds) Environmental impact assessment. Saxon House, Farnborough, UK, pp 3–15

ASIS (2003) General security risk assessment. ASIS, Alexandria, Virginia. http://www.uschamber.com/sites/default/files/issues/defense/files/guidelinesgsra.pdf

Australia Standards (2004) AS/NZS 4360:2004 Risk management. Standards Australia International and Standards New Zealand, Sydney and Wellington

Australian Government (2013) Risk analysis framework 2013. Version 4 published in May 2013. Commonwealth of Australia. p 139

Aven T (2008) Risk analysis. Wiley, Oxford

Backlund F, Hannu J (2002) Can we make maintenance decisions on risk analysis results? J Qual Maint Eng 8(1):71–91

Bandyopadhyay K, Myktyn PP, Myktyn K (1999) A framework for integrated risk management in information technology. Manag Decis 37(5):437–444
Barthwal RR (2002) Environmental impact assessment. New Age International Publishers, New Delhi
Bartlett RV, Kurian PA (1999) The theory of environmental impact assessment: implicit models of policy-making. Policy and Politics 27(4):415–433
Battelle Columbus Laboratories (1979) The selection on projects for EIA. Commission of the European Communities Environment and Consumer Protection Service, Brussels
Beck M, Drennan L, Higgins A (2002) Managing E-Risk. Association of British Insurers, London
Bedford T, Cooke R (2001) Probabilistic risk analysis: foundations and methods. Cambridge University Press, Cambridge
Boehm BW, Bose PA (1994) Collaborative spiral software process model based on theory. In: Proceedings of the 3rd international conference on the Software Process. IEEE Computer Society, Washington, D.C.
Bond AJ, Morrison-Saunders A (2011) Re-evaluating sustainability assessment: aligning the vision and the practice. Environ Impact Assess Rev 31(1):1–7
Bond A, Pope J (2012) The state of the art of impact assessment in 2012. Impact Assess Proj Apprais 30(1):1–4
Bond A, Morrison-Saunders A, Pope J (2012) Sustainability assessment: the state of the art. Impact Assess Proj Apprais 30(1):56–66
Bond A, Morrison-Saunders A, Howitt R (2013) Framework for comparing and evaluating sustainability assessment practice. In: Bond A, Morrison-Saunders A, Howitt R (eds) Sustainability assessment: pluralism, practice and progress. Taylor and Francis, London
Boroush M (1998) Understanding risk analysis: a short guide for health, safety, and environmental policy making. American Chemical Society and Resources for the Future, Washington, D.C.
Bravante MA, Holden WH (2009) Going through the motions: the environmental impact assessment of nonferrous metals mining projects in the Philippines. Pac Rev 22(4):523–547
Briggs S, Hudson MD (2013) Determination of significance in ecological impact assessment: past change, current practice and future improvements. Environ Impact Assess Rev 38:16–25
Brookes A (2001) Shared and integrative methods: environmental risk assessment and risk management. In: Morris P, Therivel R (eds) Methods of environmental impact assessment, 2nd edn. Spon Press, London, pp 351–364
Bruckner RM, List B, Schiefer J (2001) Risk management for data warehouse systems. Lect Notes Comput Sci 2114:219–229
Bujoreanu IN (2012) Risk analysis series part one—why risk analysis? J Def Resour Manage 3 (1):139–144
CAC (2006) Codex Alimentarius Commission Procedural Manual. 16th edn. Joint FAO/WHO Food Standards Programme. Available at ftp.fao.org/codex/Publications/ProcManuals/Manual_16e.pdf
Canter LW (1996) Environmental impact assessment, 2nd edn. McGraw-Hill, New York
Canter LW (1998a) Methods for effective environmental information assessment (EIA) practice. In: Porter AL, Fittipaldi JJ (eds) Environmental Methods review: retooling impact assessment for the new century. The Press Club, Fargo
Canter LW (1998b) Integration of HIA and EIA. In: Barker A, Jones C (ed) EIA Newsletter, vol 16. University of Manchester, School of Planning and Landscape, EIA Centre, Manchester, pp 10–11
Canter LW (1999) Environmental impact assessment. In: Liu DHF, Lipták BG (eds) Environmental Engineers' Handbook. CRC Press, Florida
Canter L, Ross B (2010) State of practice of cumulative effects assessment and management: the good, the bad and the ugly. Impact Assess Proj Apprais 28(4):261–268
Cashmore M (2004) The role of science in environmental impact assessment: process and procedure versus purpose in the development of theory. Environ Impact Assess Rev 24(4):403–426

Cashmore M, Gwilliam R, Morgan R, Cobb D, Bond A (2004) The interminable issue of effectiveness: substantive purposes, outcomes and research challenges in the advancement of environmental impact assessment theory. Impact Assess Proj Apprais 22(4):295–310

Cashmore M, Bond A, Cobb D (2008) The role and functioning of environmental assessment: theoretical reflections upon an empirical investigation of causation. J Environ Manage 88(4):1233–1248

Cashmore M, Bond A, Sadler B (2009) Introduction: the effectiveness of impact assessment instruments. Impact Assess Proj Apprais 27(2):91–93

Cashmore M, Richardson T, Hilding-Ryedvik T, Emmelin L (2010) Evaluating the effectiveness of impact assessment instruments: theorising the nature and implications of their political constitution. Environ Impact Assess Rev 30(6):371–379

Catchpole S, Moreno R (2012) Environmental risk and natural hazards in EIA's IAIA12 Conference Proceedings' Energy Future The Role of Impact Assessment 32nd Annual Meeting of the International Association for Impact Assessment, Centro de Congresso da Alfândega, Portugal, 27 May–1 June 2012

Chapman R (2007) Successful management of IT risk. http://www.insight.co.uk/files/other/successful%20Management%20%of%20IT%20risk.pdf

City Office of Navotas (2009) EIA of 4.0-km coastal dike, detention pond with pumping station and incidental reclamation. Philippines

Cloquell-Ballester VA, Monterde-Díaz R, Cloquell-Ballester V-A, Santamarina-Siurana M-C (2007) Systematic comparative and sensitivity analyses of additive and outranking techniques for supporting impact significance assessments. Environ Impact Assess Rev 27(1):62–83

College of Engineering and Engineering Technology Northern Illinois University (2006) Risk analysis methodologies. p 7. www.ceet.niu.edu/depts/tech/asse/tech482/risk_analysis_methodologies.doc

Cornford S (1998) Managing risk as a resource using the defect detection and prevention process. In: International conference on Probabilistic Safety Assessment and Management, pp 13–18

Covello V, Mumpower J (1985) Risk analysis and risk management: an historical perspective. Risk Anal 5(2):103–120

DEAT (2002) Ecological Risk Assessment, Integrated Environmental Management, Information Series 6. DEAT, Pretoria

DEAT (2006) Risk Management, Integrated Environmental Management Information Series 23. DEAT, Pretoria

Demidova O, Cherp A (2005) Risk assessment for improved treatment of health considerations in EIA. Environ Impact Assess Rev 25(4):411–429

Department of Health (2010) Health Risk Assessment (Scoping) Guidelines. A health risk assessment process for risk assessors for use within the scoping stages of environmental and health impact assessments. Government of Western Australia, Australia, p 23

Department of Public Works and Highways (DPWH) (1998) Environmental impact statement: the Pasig-Marikina river channel improvement project. DPWH, Philippines

DESEIN (2014) Environmental impact assessment. http://www.desein.com/Services/em-division.php

Duinker PN, Greig LA (2007) Scenario analysis in environmental impact assessment: improving explorations of the future. Environ Impact Assess Rev 27(3):206–219

Dziubiński M, Frątczaka M, Markowskib AS (2006) Aspects of risk analysis associated with major failures of fuel pipelines. J Loss Prev Process Ind 19(5):399–408

Eduljee G (1999) Risk assessment. In: Petts J (ed) Handbook on environmental impact assessment, vol 1. Blackwell Science, London, pp 374–404

Elling B (2009) Rationality and effectiveness: does EIA/SEA treat them as synonyms? Impact Assess Proj Apprais 27(2):121–131

Engel J, Mihok J, Bosak M, Majerník M (2006) Basements of crises management of environmentally dangerous zones (in Slovak). TUKE, Košice

Enserink B (2000) A quick scan for infrastructure planning: screening alternatives through interactive stakeholder analysis. Impact Assess Proj Apprais 18(1):15–22

EPA (2009) Review of the environmental impact assessment process in Western Australia. Environmental Protection Authority, Perth, Western Australia

Esteves AM, Franks D, Vanclay F (2012) Social impact assessment: the state of the art. Impact Assess Proj Apprais 30(1):34–42

European Commission (2009) Report from the Commission to the Council, the European Parliament, the European Economic and Social Committee and the Committee of the Regions. On the application and effectiveness of the EIA Directive (Directive 85/337/EEC, as amended by Directives 97/11/EC and 2003/35/EC), COM(2009) 378 final. European Commission, Brussels

Everdij MHC, Blom HAP (2013) Safety methods database. p 237. http://www.nlr.nl/downloads/safety-methods-database.pdf

FAA (2000) FAA system safety handbook. p 506. http://www.atcvantage.com/docs/FAA_System_Safety_Handbook_2001.pdf

Fairman R, Mead CD, Williams WP (1999) Environmental risk assessment—approaches, experiences and information sources. Monitoring and Assessment Research centre, King's College, London. Published by European Environment Agency—EEA Environmental issue report No 4

FAO (2009) Environmental impact assessment and monitoring in aquaculture. Requirements, practices, effectiveness and improvements. FAO, Rome, p 675. ftp.fao.org/docrep/fao/012/i0970e/i0970e.pdf

FAO, WHO (2006) Food safety risk analysis: a guide for national food safety authorities. FAO Food and Nutrition paper 87, Rome

Fera M, Macchiaroli R (2010) Appraisal of a new risk assessment model for SME. Saf Sci 48 (10):1361–1368

Fitzpatrick P (2006) In it together: organizational learning through participation in environmental assessment. J Environ Assess Policy Manage 8(2):157–182

Flanagan R, Norman G (1993) Risk management and construction. Blackwell Publication, London

Frosdick S (1997) The techniques of risk analysis are insufficient in themselves. Disaster Prev Manage 6(3):165–177

Fundingsland Tetlow M, Hanusch M (2012) Strategic environmental assessment: the state of the art. Impact Assess Proj Apprais 30(1):15–24

GESAMP (2008) (IMO/FAO/UNESCO-IOC/UNIDO/WMO/IAEA/UN/UNEP Joint Group of Experts on Scientific Aspects of Marine Environmental Protection). Assessment and communication of environmental risks in coastal aquaculture. FAO, Reports and Studies GESAMP (76), Rome

Gilbuena R, Kawamura A, Medina R, Amaguchi H, Nakagawa N, Bui DD (2013) Environmental impact assessment of structural flood mitigation measures by a rapid impact assessment matrix (RIAM) technique: a case study in Metro Manila, Philippines. Sci Total Environ 456–457:137–147

Gough JD (1988) Review of environmental impact assessment and risk assessment procedures. Unpublished report for the Ministry for the Environment, Centre for Resource Management, University of Canterbury and Lincoln College, Christchurch

Gough JD (1989) A strategic approach to the use of environmental impact assessment and risk assessment within the decision-making process. Information Paper no. 13. Centre for Resource Management. University of Canterbury and Lincoln College

Graham JD (1995) Historical perspective on risk assessment in the federal government. Toxicology 102(1–2):29–52

Greiving S (2005) Impact Assessment as procedural framework for risk governance. Paper presentation at the final conference of the IMP3 Project "Improving European EIA-Implementation", Federal Ministry for Agriculture, Forestry, Environment and Water Management, Austria, 22 November 2005

Grima AP, Timmerman P, Fowle CD, Byer P (1986) Risk management and EIA: research needs and opportunities. Canadian Environmental Assessment Research Council, Quebec

Halliday S, Badenhorst K, Solms R (1996) A business approach to effective information tech-nology risk analysis and management. Inf Manage Comput Secur 4(1):19–31
Harding R (1998) Uncertainty, risk and the precautionary principle. In: Harding R (ed) Environmental decision-making. The roles of scientists, engineers and the public. The federation press, Sydney, pp 164–191
Harris-Roxas B, Viliani F, Bond A, Cave B, Divall M, Furu P, Harris P, Soeberg M, Wernham A, Winkler M (2012) Health impact assessment: the state of the art. Impact Assess Proj Apprais 30(1):43–52
Harrop DO, Nixon JA (1999) Environmental assessment in practice. Routledge, London
Higuera RP, Haimes YY (1996) Software risk management. Technical Report CMU/SEI-96-TR-012, ESC-TR-96-012, Software Engineering Institute, Carnegie Mellon University, Pittsburgh, PA, USA
Hill RA (2005) Conceptualizing risk assessment methodology for genetically modified organisms. Environ Biosafety Res 4(2):67–70
IEMA (Institute of Environmental Management and Assessment) (2011) The state of environmental impact assessment practice in the UK. IEMA, Lincoln, UK
Ijäs A, Kuitunen MT, Jalava K (2010) Developing the RIAM method (rapid impact assessment matrix) in the context of impact significance assessment. Environ Impact Assess Rev 30(2):82–89
ISO/IEC Guide 73: (2009) Risk management—Vocabulary, ISO/IEC
IUCN (2000) IUCN guidelines for the prevention of biodiversity loss due to biological invasion. The World Conservation Union (approved by the IUCN Council, February 2000)
JaE (2009) Access to justice in Slovakia. In: Toolkits A (ed) Access to justice in environmental matters. Justice and Environment—European Association of Environmental Law Organisations, Brno, pp 1–25
Jay S, Jones C, Slinn P, Wood C (2007) Environmental impact assessment: retrospect and prospect. Environ Impact Assess Rev 27(4):287–300
Jha-Thakur U, Gazzola P, Peel D, Fischer TB, Kidd S (2009) Effectiveness of strategic environmental assessment-the significance of learning. Impact Assess Proj Apprais 27(2):133–144
Johansson H, Jönsson H (2007) Methods for risk analysis—system perspectives (in German). Lunds universitet, Lund
Jurison J (1999) Software project management: a manager's view. Commun Assoc Inf Syst 2 (17):1–57
Kandráč J, Skarba D (2000) Methodical approach for risk assessment of dangerous enterprises and study about companies in the Slovak republic (in Slovak). Risk Consult, Bratislava
Kenis M, Bacher S, Baker RHA, Branquart E, Brunel S, Holt J, Hulme PE, MacLeod A, Pergl J, Petter F, Pyšek P, Schrader G, Sissons A, Starfinger U, Schaffner U (2012) New protocols to assess the environmental impact of pests in the EPPO decision-support scheme for pest risk analysis. Bull OEPP/EPPO Bull 42(1):21–27
Kirkpatrick C, George C (2006) Methodological issues in the impact assessment of trade policy: experience from the European Commission's Sustainability Impact Assessment (SIA) programme. Impact Assess Proj Apprais 24(4):325–334
Kjorven O (1998) Environmental risk assessment in development programmes: the experience of the World Bank. In: Calow P (ed) Handbook of environmental risk assessment and management. Blackwell Science, Oxford, UK, pp 506–519
Kontio J (1997) The riskit method for software risk management, version 1.00. University of Maryland. http://www.soberit.hut.fi/T-76.115/02-03/palautukset/groups/pmoc/de/riskit.pdf
Kozová M, Drdoš J, Pavličková K, Úradníček Š (1996) Environmental impact assessment. Comenius University in Bratislava, Bratislava
Kruopienė J, Židonienė S, Dvarionienė J (2009) Current practice and shortcomings of EIA in Lithuania. Environ Impact Assess Rev 29(5):305–309
Lawrence DP (1997) The need for EIA theory-building. Environ Impact Assess Rev 17(2):79–107
Lawrence DP (2000) Planning theories and environmental impact assessment. Environ Impact Assess Rev 20(6):607–625

Lawrence DP (2007a) Impact significance determination—designing an approach. Environ Impact Assess Rev 27(8):730–754
Lawrence DP (2007b) Impact significance determination—pushing the boundaries. Environ Impact Assess Rev 27(8):770–788
Leknes E (2001) The roles of EIA in the decision-making process. Environ Impact Assess Rev 21(4):309–334
Lexer W, Paluchova K, Schwarzl B (2006) Risk assessment. IMProving the IMPlementation of Environmental IMPact Assessment, "(IMP)3". Risk Assessment D 3.2 Report WP 3. Österreichisches Institut für Raumplanung, Vienna. http://www.umweltbundesamt.at/fileadmin/site/umweltthemen/UVP_SUP_EMAS/IMP/IMP3-Risk_Assessment.pdf
Lohani B, Evans JW, Everitt RR, Ludwig H, Carpenter RA, Tu SL (1997) Environmental impact assessment for developing countries in Asia volume 1—overview. Asian Development Bank. http://www.adb.org/sites/default/files/pub/1997/eia-developing-countries-asia.pdf. Accessed Jan 2012
Luciak M (2012) Application of the Act on Environmental Impact Assessment (in Slovak) In: ELSEWA Ltd. (ed) Proceeding from 2nd international conference: Environment—problems and solutions; air—water—soil, Košice, pp 12–17
Luciak M, Nižňanský G (2013) Environmental Impact Assessment. In: ELSEWA Ltd. (ed) Proceeding from 2nd international conference: Environment—problems and solutions; air—water—soil, Košice, pp 5–20
Ludwig HF, Gunartnam D, Yuming S (1995) Environmental impact assessment for Xiaolangdi Yellow River multi-purpose economic-cum-environmental improvement project. Environmentalist 15:45–57
Mandelik Y (2005) Issues and dilemmas in ecological scoping: scientific, procedural and economic perspectives. Impact Assess Proj Apprais 23(1):55–63
Marara M, Okello N, Kuhanwa Z, Douven W, Beevers L, Leentvaar J (2011) The importance of context in delivering effective EIA: case studies from East Africa. Environ Impact Assess Rev 31(3):286–296
Marhavilas PK, Koulouriotis D, Gemeni V (2011) Risk analysis and assessment methodologies in the work sites: on a review, classification and comparative study of the scientific literature of the period 2000–2009. J Loss Prev Proc Ind 24(5):477–523
Marshall R, Arts J, Morrison-Saunders A (2005) International principles for best practice EIA follow-up. Impact Assess Proj Apprais 23(3):175–181
McDaniels T, Small M (2004) Risk analysis and society: an interdisciplinary characterization of the field (ed). Cambridge University Press, Cambridge
McVicar AH (2004) Management actions in relation to the controversy about salmon lice infections in fish farms as a hazard to wild salmonid populations. Aquacult Res 35(8):751–758
Mobey A, Parker D (2002) Risk evaluation and its importance to project implementation. Work study 51(4):202–208
MoE (2010) State of the Environment of the Slovak Republic in 2009 (in Slovak). Ministry of Environment/Slovak Environmental Agency, Bratislava/Banská Bystrica
MoE, SEA (2012) Conclusions from 2nd international conference SEA/EIA 2012 (in Slovak). Ministry of Environment/Slovak Environmental Agency, Bratislava/Banská Bystrica, Tále. http://www.enviroportal.sk/clanok/eiasea-zhodnotenie-konferencie. Accessed 17 July 2012
MoE (2013) State of the Environment of the Slovak Republic in 2012 (in Slovak). Ministry of Environment/Slovak Environmental Agency, Bratislava/Banská Bystrica
Morgan RK (1998) Environmental impact assessment: a methodological perspective. Kluwer Academic, Dordrecht
Morgan RK (2012) Environmental impact assessment: the state of the art. Impact Assess Proj Apprais 30(1):5–14
Morrison-Saunders A, Arts J (2004) Assessing impact: handbook of EIA and SEA follow-up. Earthscan James & James, London
Morrison-Saunders A, Bailey M (2009) Appraising the role of relationships between regulators and consultants for effective EIA. Environ Impact Assess Rev 29(5):284–294

Mosler D (2006) Risk assessment of dams by FMECA (in Czech). In: Student scientific and practice activity. 2006. VUT FAST, Brno
Mullai A (2006) Risk management system—risk assessment frameworks and techniques. DaGoB Project Office, Turku
Munier N (2004) Multicriteria environmental assessment, a practical guide. Kluwer Academic Publishers, London
Mustow SE, Burgess RF, Walker N (2005) Practical methodology for determining the significance of impacts on the water environment. Water Environ J 19(2):100–108
National Academy of Sciences (2011) Improving health in the United States: the role of health impact assessment. The National Academies Press, Washington, D.C.
National Council of the Slovak Republic (2005) Act of Law No. 24/2006 from December 14th 2005 on Environmental Impact Assessment and on amendments to certain acts
Natural Resources Commission (2013) Risk management policy and framework. Document No: D12/4540. http://www.nrc.nsw.gov.au/content/documents/Policy%20-%20Risk%20management%20-%20Mar%202013.pdf
N.E.M. Business Solutions (2002) Risk analysis methodologies. N.E.M Business Solutions, United Kingdom. www.cip.ukcentre.com/risk.htm
Nijkamp P (1980) Environmental policy analysis. Operational methods and models. Wiley, Chichester
O'Riordan T, Sewell WRD (1981) From project appraisal to policy review. In: O'Riordan T, Sewell WRD (ed) Project appraisal and policy review. Wiley, Chichester
Ortolano L, Shepherd A (1995) Environmental impact assessment: challenges and opportunities. Impact Assess 13(1):3–30
Pačaiová H, Sinay J, Glatz J (2009) Security and risks of technical systems (in Slovak). TUKE, Košice
Paustenbach DJ (1995) Retrospective on U.S. health risk assessment. How others can benefit. Risk: Health Saf Environ 6:283–332
Perry JG, Hayes RW (1985) Risk and its management in construction projects. Proceedings of the Institution of Civil Engineers, Part 1, v78, Design and Construction, Engineering Management Group, pp 499–521
Petts J (1997) Environmental assessment and risk assessment, teaching-materials prepared for the course Environmental Assessment and Audit as part of the collaborative programme on Environmental Impact Assessment and Project Appraisal in Developing Countries at the Environmental Impact Assessment Center, School of Planning and Landscape, Manchester
Petts J (1999) Environmental impact assessment—overview of purpose and process. In: Petts J (ed) Handbook of environmental impact assessment, vol 1. Blackwell Science, London, pp 3–11
Petts J, Eduljee G (1994) Environmental impact assessment for waste treatment and disposal facilities. Wiley, Chichester
Phillips MJ, Subasinghe RP (2008) Application of risk analysis to environmental issues in aquaculture. In: Bondad-Reantaso MG, Arthur JR, Subasinghe RP (ed) Understanding and applying risk analysis in aquaculture. FAO Fisheries and Aquaculture Technical Paper. No. 519. Rome, FAO. pp 101–119
Pinho P, McCallum S, Santos Cruz S (2010) A critical appraisal of EIA screening practice in EU Member States. Impact Assess Proj Apprais 28(2):91–107
Pope J, Annandale D, Morrison-Saunders A (2004) Conceptualising sustainability assessment. Environ Impact Assess Rev 24(6):595–616
Pope J, Bond A, Morrison-Saunders A, Retief F (2013) Advancing the theory and practice of impact assessment: setting the research agenda. Environ Impact Assess Rev 41:1–9
Poulard C, Lafont M, Lenar-Matyas A, Łapuszek M (2010) Flood mitigation designs with respect to river ecosystem functions—a problem oriented conceptual approach. Ecol Eng 36(1):69–77

Rainer RK Jr, Snyder ChA, Carr HH (1991) Risk analysis for information technology. J Manage Inf Syst 8(1):129–147
Rajaram T, Das A (2011) Screening for EIA in India: enhancing effectiveness through ecological carrying capacity approach. J Environ Manage 92(1):140–148
Rasche T (2001) Risk analysis methods—a brief review. The University of Queensland. Mineral Industry Safety and Health Center. http://www.mishc.uq.edu.au/Publications/Risk_Analysis_Methods_a_Brief_Review.pdf
Raspotnig Ch, Opdahl A (2013) Comparing risk identification techniques for safety and security requirements. J Syst Softw 86(4):1124–1151
Rechard RP (1999) Historical relationship between performance assessment for radioactive waste disposal and other types risk assessment. Risk Anal 19(5):763–807
Reniers GLL, Dullaert W, Ale BJM, Soudan K (2005) Developing an external domino prevention framework: Hazwim. J Loss Prev Process Ind 18(3):127–138
Retief F (2010) The evolution of environmental assessment debates: critical perspectives from South Africa. J Environ Assess Policy Manage 12(4):375–397
Říha J (1987) Multicriteria assessments of investments projects (in Czech). SNTL, Praha
Říha J (2001) Environmental impact assessment. Methods for preliminary decision analysis (in Czech). ČVUT, Praha
Říha J (1995) Environmental impact assessment of investments. Multicriteria analysis and EIA (in Czech). Academia, Praha
Rouvroye JL, van den Bliek EG (2002) Comparing safety analysis techniques. Reliab Eng Syst Saf 75(3):289–294
Sadler B (1996) Environmental assessment in a changing world: evaluating practice to improve performance. Final Report of the International Study of the Effectiveness of Environmental Assessment. Canadian Environmental Assessment Agency, Canada
Sadler B et al (2011) Taking stock of SEA. In: Sadler B (ed) Handbook of strategic environmental assessment. Earthscan, London, pp 1–18
Sánchez LE, Morrison-Saunders A (2011) Learning about knowledge management for improving environmental impact assessment in a government agency: the Western Australian experience. J Environ Manage 92(9):2260–2271
Schierow LJ (2004) The role of risk analysis and risk management in environmental protection, CRS Issue Brief for Congress, Congressional Research Service, Library of Congress, Washington, D.C.
Segudovic H (2006) Qualitative risk analysis method comparison. Infigo IS d.o.o, Zagreb
Shine C, Williams N, Gündling L (2000) A guide to designing legal and institutional frameworks on alien invasive species. International Union for Conservation of Nature, Switzerland
Šimák L (2006) Risk management (in Slovak). ZU, Žilina
Smejkal V, Rais V (2013) Risk management in firms and other companies (in Czech). Expert. Grada Publishing, Praha
Smith PG, Merritt GM (2002) Proactive risk management: controlling uncertainty in product development. Productivity Press, New York, p 246
Snell T, Cowell R (2006) Scoping in environmental impact assessment: balancing precaution and efficiency? Environ Impact Assess Rev 26(4):359–376
Sommerville I (2010) Software engineering. 9th ed. Addison-Wesley, USA, p 790. http://poincare.matf.bg.ac.rs/~vladaf//Courses/Matf%20RS2/Text/Software%20Engineering%209ed.pdf
SRA (2004) Glossary of risk analysis terms. McLean, VA. http://www.sra.org/sites/default/files/docs/SRA_Glossary.pdf
Srinivas H, Nakagawa Y (2008) Environmental implications for disaster preparedness: lessons learnt from the Indian Ocean tsunami. J Environ Manage 89(1):4–13
Startienė G, Remeikienė R (2007) Methodology of business risk analysis and its practical application in the enterprises working in the global market. Eng Econ 3:1–15

Steinemann A (2001) Improving alternatives for environmental impact assessment. Environ Impact Assess Rev 21(1):3–21

Stewart JMP, Sinclair AJ (2007) Meaningful public participation in environmental assessment: perspectives from Canadian participants, proponents, and government. J Environ Assess Policy Manage 9(2):161–183

STN 01 0380: (2003) Risk management (AS/NZS 4360:1999). Bratislava

STN ISO 31000: (2011) Risk management. Principles and guidelines. SÚTN, Bratislava

Stoneburner G, Goguen A, Feringa A (2002) Risk management guide for information technology systems recommendations of the National Institute of Standards and Technology. http://csrc.nist.gov/publications/nistpubs/800-30/sp800-30.pdf

Strutt J (1993) Risk management. Unpublished paper, Centre for Industrial Safety and Reliability, Cranfield University

Suh B, Han I (2002) The IS risk analysis based on a business model. Inf Manag 41(2):149–158

Tamura H, Fujita S, Hiroshi K (1994) Decision analysis for environmental impact assessment and consensus formation among conflicting multiple agents—including case studies for road traffic. Sci Total Environ 153(3):203–210

Taylor CN, Bryan CH, Goodrich CG (2004) Social assessment: theory, process and techniques, 3rd edn. Ecology Press, Wisconsin

The E7 Network of Expertise for the Global Environment (1997) Environmental impact assessment. An electric utility overview. The E7 Network of Expertise for the Global Environment, Canada. http://www.e8.org/upload/File/E7_EIA_Manual_EN.pdf

Tichý M (2006) Risk control: analysis and management. C. H, Beck, Prague

Treweek J (1999) Ecological impact assessment. Blackwell Science, Oxford

USEPA (1992) Framework for ecological risk assessment. EPA Risk Assessment Forum. EPA/630/R-92/001

USEPA (1998) Guidelines for ecological risk assessment. US Environmental Protection Agency, Risk Assessment Forum, Washington D.C., EPA/630/R-95-002F

Vanclay F, Bronstein DA (1995) Environmental and social impact assessment. Wiley, Chichester

Vaughan EJ (1997) Risk management. Wiley, New York

Waldeck S, Morrison-Saunders A, Annandale D (2003) Effectiveness of non-legal EIA guidance from the perspective of consultants in Western Australia. Impact Assess Proj Apprais 21 (3):251–256

Walker LJ, Johnston J (1999) Guidelines for the assessment of indirect and cumulative impacts as well as impact interactions. EC DG XI, Environment, Nuclear Safety & Civil Protestion, Luxembourg

Wang YM, Yan JB, Xu DL (2006) Environmental impact assessment using evidential reasoning approach. Eur J Oper Res 174(3):1885–1913

Wathern P (1988) An introductory guide to EIA. In: Wathern P (ed) EIA: theory and practice. Unwin Hyman, London

WB (1991) Environmental assessment sourcebook. World Bank Technical Paper 154. Washington, D.C.

Welsh T (2012) Full Monte—The better approach to Schedule Risk Analysis. User guide. Tony Welsh, Barbecana

Weston J (2011) Screening for environmental impact assessment projects in England: what screening? Impact Assess Proj Apprais 29(2):90–98

White D (1995) Application of systems thinking to risk management: a review of the literature. Manag Decis 33(10):35–45

Wood Ch (2003) Environmental impact assessment: a comparative review, 2nd edn. Prentice Hall, Harlow

Zeleňáková M (2009) Flood risk assessment (in Slovak). TUKE, Košice

Zelený J (2006) Analysis, assessment and evaluation of risks, principle of duality and probability (in Slovak). TUZVO, Zvolen

Zhou K, Sheate WR (2011) EIA application in China's expressway infrastructure: clarifying the decision-making hierarchy. J Environ Manage 92(6):1471–1483
Zvijáková L (2013) Aplikácia rizikovej analýzy v procese posudzovania vplyvov (vybraných stavieb) na životné prostredie. [The application of risk analysis in the environmental impact assessment (selected constructions)]. Dissertation, Technical university of Košice
Zvijáková L, Zeleňáková M, Purcz P (2014) Evaluation of environmental impact assessment effectiveness in Slovakia. Impact Assess Proj Apprais 32(2):150–161

Chapter 2
Risk Analysis in Impact Assessment

While there are some studies in the literature considering risk analysis in construction projects (Zavadskas et al. 2010), studies of risk analysis regarding water management constructions, especially assessment of flood protection structures (FPS), are very limited. In classical project risk-analysis techniques, risk-rating values are calculated by multiplying probability and impact values, but direct analysis of the linguistic factors involved is often neglected (Dikmen et al. 2007). This book introduces a new approach to risk assessment of activities in water management (FPS projects) using risk analysis.

Various approaches for integrating risk analysis into the EIA process have been suggested (e.g., Kwiatkowski 1998; AGIP KCO 2004; Demidova and Cherp 2005; and Catchpole and Moreno 2012). Despite its growing acceptance, no reference has been found in the present study to its use in EIA applied to FPS worldwide. The applicability of risk analysis in the Slovakian EIA system has also yet to be tested. In our opinion this country could benefit from trying out the risk-analysis method, and we understand the importance of providing examples of its application. At the same time though, the risk-analysis technique must conform to the general impact-assessment approach on which the Slovakian EIA system is abused.

This chapter describes the design adopted by this research to achieve the aims and objectives stated in the introduction. Section 2.1 discusses the methodology to be used in the study, the stages by which the methodology will be implemented, and the research design; Sect. 2.2 gives details of establishing the context of the study; Sect. 2.3 describes the risk-analysis method used in the study; and Sect. 2.4 outlines the procedure used for decision-making.

M. Zeleňáková and L. Zvijáková, *Using Risk Analysis for Flood Protection Assessment*, DOI 10.1007/978-3-319-52150-3_2

2.1 Methodology and Research Design

2.1.1 Methodology

Floods are the most frequent natural disaster worldwide and a major natural hazard in Europe in terms of social and economic impacts. In the last 15 years, Europe has suffered over 100 major damaging floods which have caused, in total, more than 1000 casualties, affected more than 3.4 million people, about half a million of whom have been displaced, and at least € 25 billion in insured economic losses. Additionally, floods cause important environmental impacts since they seriously affect the quality of water sources and can distribute large amounts of sediments and pollutants (Papa and Torres 2012). The aim of this book is to develop a methodology for the analysis and evaluation of environmental impacts of proposed activities—flood protection structures using a risk-analysis method. The proposed methodology has been applied to one proposed flood protection structure (Zvijáková 2013). The application of developed methodology for the EIA process will produce indications for improvements, or for more effective implementation and performance of this process. The main objective of the book is to improve the course of the EIA process. More specifically, the methodology's objectives are to:

- establish the principles and methods of risk analysis in the EIA process,
- support the comparison of variants of the proposed activity on the basis of their evaluation through the concept of "risk", while strengthening the decision-making processes within the EIA process,
- increase knowledge on emerging risks and ensure their monitoring,
- create a new methodological approach applicable to the EIA process,
- enable interested parties to implement the proposed methodology consistently and effectively as part of assessment reports within the EIA process.

The authors have determined that a risk-based approach may be applied in the EIA process in Slovakia. It is assumed that this process will be applied during the scoping phases of the EIA and will include consideration of potential impacts of developments on the environment and humans.

To integrate risk analysis within EIA, it is appropriate that the criteria used within the EIA risk-based approach are consistent with the terminology and understandings used within the water-management sector.

This book provides a framework for the risk analysis component mainly of the scoping phase within the EIA process (see Fig. 2.1).

The proposed methodology consists of three stages, which includes a number of key elements and activities.

Element—the process of risk analysis according to the proposed methodology consists of four activities: creation of a set of risk factors (*A*–*Z*), determining the relative importance (weight) of the risk factors (w_i), creation of risk criteria for risk factors and determination of criterion scores (0.2–1.0).

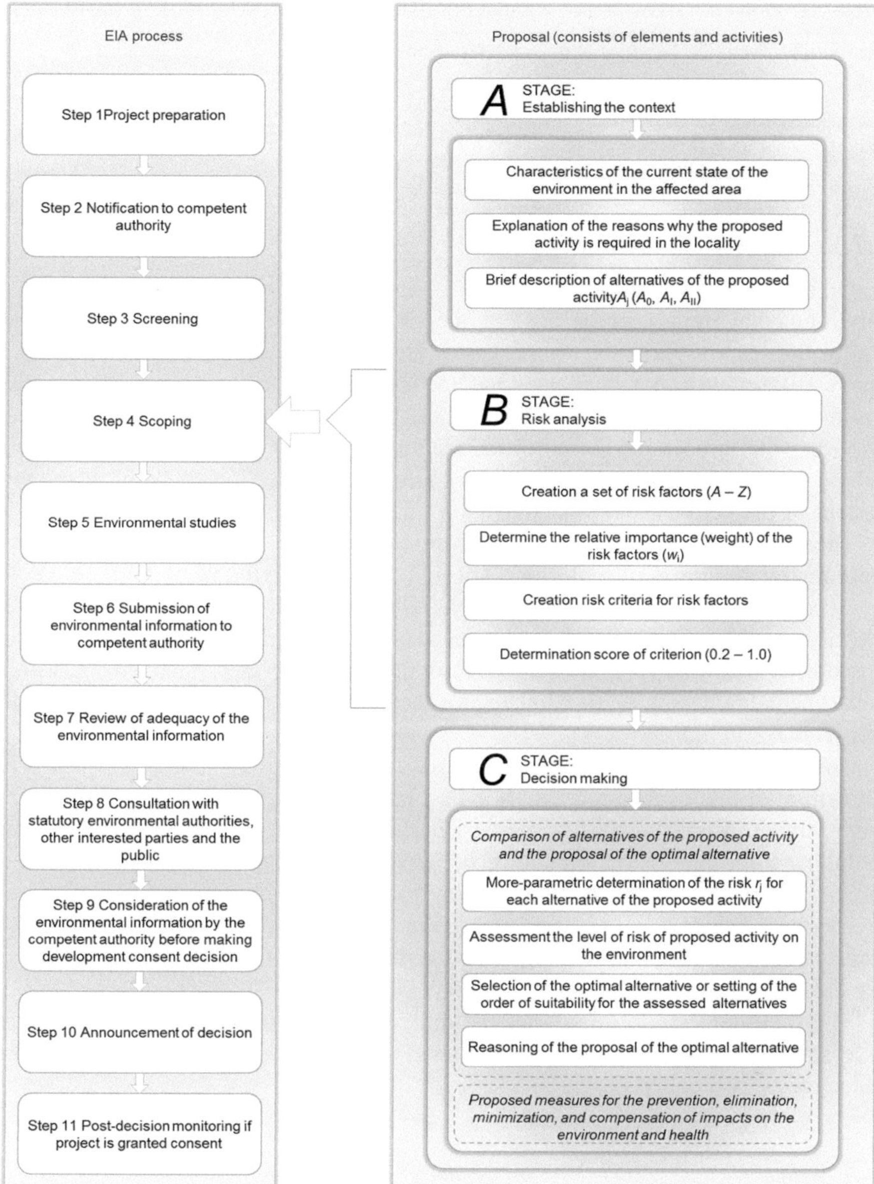

Fig. 2.1 Flow chart of the EIA process (EC 2001) and proposal methodology (Zvijáková 2013)—integration of risk analysis (RA) into the environmental impact assessment (EIA) process

Table 2.1 Proposed activity which is subject to EIA (National Council of the Slovak Republic)

Item no.	Activity, facilities and installations	Threshold values	
		Part A (compulsory assessment)	Part B (screening procedure)
7.	Flood protection objects		Without limit

2.1.2 Research Design

Environmental impact assessment of the proposed activity is a standard output within the responsibility of the proposer. Our methodical procedure introduces a quantitative approach which uses the method of risk analysis known as multi-parametric expression of risk. The role of the method is to unify the means of assessing the impacts of the proposed activity on the environment and the establishment of maximum standardized procedure for the selection of the most suitable variant of the activity and objectivization of the EIA process.

The proposed methodology of environmental impact assessment of flood protection structures is applicable for a specific type of activity.

According to Annex no. 8 to Law No. 24/2006 Coll. (National Council of the Slovak Republic 2005), flood protection structures are included in the list of proposed activities which are subject to EIA in field no 10: Water Management, whose departmental authority is the Ministry for the Environment of the Slovak Republic (see Table 2.1).

The basic principle of the methodology is to calculate the risk index, which is an estimation of the level of risk that flood protection structures may represent for the environment. Risk analysis is based on the principle that every construction, including flood-protection structures, not only by its technical character adversely affects the hydrological, morphological and geographical, ecological, archaeological/historical, and socio-economic characteristics of the area. Negative impacts can be quantified by calculating the risk index of flood-protection structures for the environment.

The following section describes the different elements and activities of impact assessment of flood protection structures which are designed to achieve the objective of risk analysis.

2.2 Establishing the Context

The primary step in the initiation of the impact assessment of the proposed activity is a comprehensive understanding of the variants of the proposed activity. It is important to know the characteristics of the current state of the environment, which

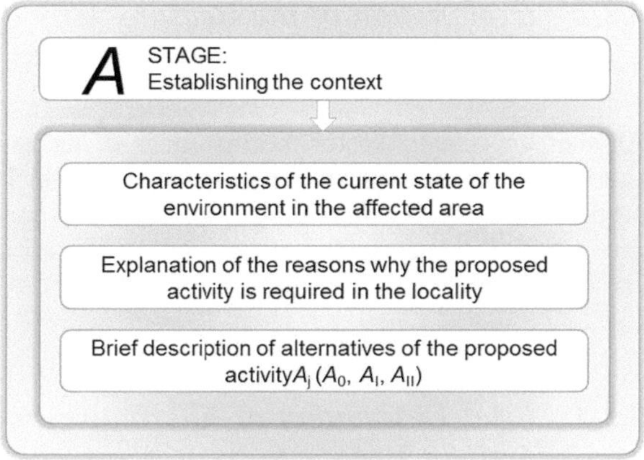

Fig. 2.2 The first stage of impacts assessment of the proposed activity and its activities

is the main task in defining the context. To complete this stage, it is necessary to perform the procedure as outlined in Fig. 2.2.

2.2.1 *Characteristics of the Current State of the Environment in the Affected Area*

For further steps in the procedure, it is necessary to know the current state of the environment in the area where the proposed activity is to be carried out.

The impact of the proposed activity on the environment is always limited to an area where the effects of the activity are immediately evident on site, which may be affected by visual, auditory, olfactory or other factors. Such an area is designated as the affected area.

Basic information about the present state of the environment in the affected area, according to Annex no. 9 to Slovakian Law No. 24/2006 Coll. (National Council of the Slovak Republic 2005), consists of the following:

- *"characteristics of the natural environment including protected areas [e.g., proposed protected bird areas, areas of European interest, coherent European network of protected areas (Natura 2000), national parks, protected landscape areas, protected water-management areas]";*
- *"landscape, landscape character, stability, protection, scenery";*
- *"population, its activities, infrastructure, cultural and historical values of the area";*
- *"current state of the quality of the environment including health".*

According to Annex no. 11 to Law No. 24/2006 Coll. in Slovakia, the description of the current state of the environment in the affected area consists of information regarding nineteen topics further specified in the attachment.

2.2.2 Explanation of the Reasons Why the Proposed Activity Is Required in the Locality

Another important step for the first stage of the evaluation process is the justification for the need of the proposed action, which is based on the aim of the proposed activity. It is necessary to know the reasons which have led to the planned activity and to understand its nature and the circumstances under which the activity can be performed.

2.2.3 Brief Description of Alternatives to the Proposed Activity A_j (A_0, A_I, A_{II})

The purpose of alternatives is "to find the most effective way of meeting the need and purpose of the proposal, either through enhancing the environmental benefits of the proposed activity and or through reducing or avoiding potentially significant negative impacts" (DEAT 2004).

Among the most important issues in the scoping phase of the EIA process is consideration of potential alternatives (DEAT 2002). Their significance is highlighted by Glasson et al. (1999) and by the Council of Environmental Quality (CEQ) in the United States, which describes the consideration of alternatives as the "heart" of EIA (CEQ 1978; Magness 1984). Considering alternatives is a critical aspect of the environmental-assessment process. Its purpose is to provide a framework for sound decision-making based on the principles of sustainable development (DEAT 2004).

Article 5(3d) of the EIA Directive requires the developer to include in the environmental information "*...an outline of the main alternatives studied by the developer and an indication of the main reasons for this choice, taking into account the environmental effects*".

Some EU member states have made consideration of alternatives a mandatory requirement for EIA, whilst others leave it to the developer to decide if alternatives are relevant to their project.

According to Section 22(3) of Law No. 24/2006, "*the preliminary environmental study must contain two alternatives of the proposed activity at least, as well as the zero alternative*". The zero alternative is the state that would remain if the proposed activity was not carried out.

According to Section 22(7) of Law No. 24/2006, "*the competent authority, on the request of the proponent, shall abstain from the requirement of an alternative solution of the proposed activity mainly in cases where no other locality is available or if no other technology for the proposed activity exists. If from the comments to the preliminary environmental study, submitted according to Section 23(5), the need to assess another real alternative of the proposed activity results, this fact will be taken into account in further proceedings according to this Law*".

Alternatives are substantially different manners in which the proponent can feasibly meet the project's objectives, for example by carrying out a different type of action, choosing an alternative location or adopting a different technology or design for the proposed project. Alternatives merge into mitigating measures where specific changes are made to the project design or to methods of construction or operation to avoid, reduce or mitigate environmental impacts (reduce significant adverse impacts).

DEAT (2004) presents the key criteria for identifying alternatives within projects when applying EIA to them. The various categories of alternatives that can be identified include: "activity alternatives; location alternatives; process alternatives; demand alternatives; scheduling alternatives; input alternatives; routing alternatives; site layout alternatives; scale alternatives; and design alternatives".

The "no project" (or zero) alternative must also be considered as the baseline against which the environmental impacts of the project should be considered. This may include changes from the present day situation as a result of other developments taking place in the vicinity and changes in environmental conditions of the study area.

The *Checklist on Alternatives and Mitigation* in Guidance on EIA Scoping (EC 2001) provides a useful list to consider when thinking about the different types of alternatives and mitigation which a developer should consider.

The key criteria for identifying appropriate alternatives are that they should be "practicable", "feasible", "relevant", "reasonable" and "viable". When a range of alternatives exists, it is necessary to identify which of them are applicable to the specific projects under consideration (DEAT 2004).

2.3 Risk Analysis

The effort to produce an empirical description of several factors simultaneously leads to multiparametric risk description (Tichý 2006). As stated in Sect. 1.2.4, this is one of the methods of risk analysis, and this book proposes an approach that uses this method. In this second phase, it is necessary to take the steps illustrated in Fig. 2.3 which is described briefly below.

2.3.1 Creation a Set of Risk Factors (A–Z)

Environmental parameters are indicators of impacts/effects of flood protection structures (Zeleňáková and Zvijáková 2012). The function of parameters is to enable relatively easily quantifiable measurement of the impact of the proposed activity on the environment. The proposed parameters are divided into five groups according to their character:

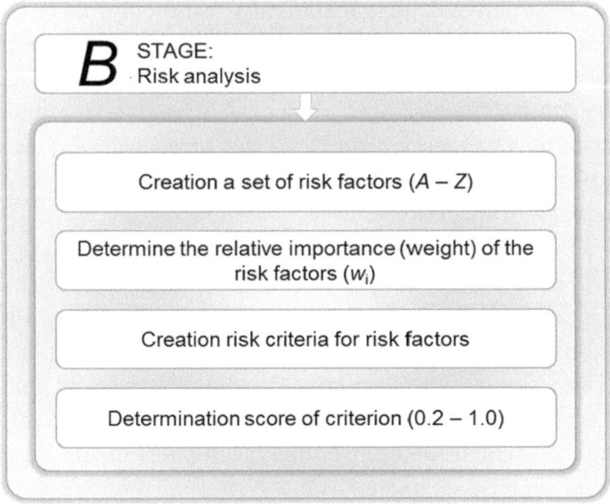

Fig. 2.3 The second stage of impact assessment of the proposed activity

- ***hydrological parameters***:

  A. maximum specific drainage;
  B. 100-year flow;
  C. designated flow;
  D. average annual rainfall;

- ***morphological and morphometric parameters***:

  E. forestation;
  F. coefficient of saturation in the basin;
  G. stream-flow character;
  H. average longitudinal-gradient flow;
  I. type of the basin;
  J. catchment area;
  K. soil type;
  L. slope of basin;

- ***ecological and historical parameters***:

  M. ecological significance of the area;
  N. vulnerability of protected species of fauna, flora and their biotopes;
  O. change to the landscape;
  P. cultural and historical importance of the territory;
  Q. vulnerability of archaeological and paleontological sites and important geological sites;

- ***territorial parameters***:

  R. permanent population in the protected area;
  S. coefficient of built-up area;
  T. type and importance of transport;
  U. infrastructure of villages;
  V. production activity of the territory;
  W. degree of environmental and human damage;
  X. total cost of the proposed activity;

- ***technical parameters***:

  Y. distance of the place of proposed-activity implementation from built-up areas;
  Z. state of existing flood protection structures.

Twenty-six risk criteria have been identified and defined in this flood-mitigation-measures proposal, based on expert interviews, field studies and literature review. The proposed alphabet of parameters is used for the calculation of risk indices of the proposed construction or flood protection structure (Zeleňáková et al. 2012).

2.3.2 Determine the Relative Importance (Weight) of the Risk Factors (w_i)

In the set of parameters, not all elements of the set Pa_i have the same relative importance in relation to the particular problem under consideration. This relative significance or importance is simply referred to as a weight parameter w_i (Říha 2001). For a summary of recommended methods for determining the weights of parameters, the criteria are clearly stated for example in Říha (2001) or Křupka et al. (2012). This scale provides information about the relative social importance (impact) of individual parameters within a given set of Pa_i (*A–Z*).

To determine the relative importance (weight) of parameters, a survey was carried out involving twenty experts in water management and professionally qualified persons who have experience in the design and implementation of flood mitigation measures.

For the purposes of this work, a direct method of determining weights is used, based on the scoring method known as Metfessel allocation (Křupka et al. 2012). This assumes that the user is able to quantitatively evaluate the importance of the parameters relating to their impact on the environment. The user evaluates the *i*th parameter with value b_i, if it lies in the scale, e.g., b_i <0, 100>. The more important the parameter is, the higher its score is. While the scoring method requires the user to provide quantitative evaluation of indicators, it also allows for a more

Table 2.2 Determination of weight of parameters Pa_i (*A–Z*) using scoring

	$Pa_1 = A$	$Pa_2 = B$	$Pa_3 = C$	$Pa_4 = D$	$Pa_5 = E$	$Pa_6 = F$	$Pa_7 = G$
Points b_i	7	5	10	9	3	3	1
Weight w_i	0.07	0.05	0.10	0.09	0.03	0.03	0.01
	$Pa_8 = H$	$Pa_9 = I$	$PR_{10} = J$	$Pa_{11} = K$	$Pa_{12} = L$	$Pa_{13} = M$	$Pa_{14} = N$
Points b_i	3	2	1	1	3	2	2
Weight w_i	0.03	0.02	0.01	0.01	0.03	0.02	0.02
	$Pa_{15} = O$	$Pa_{16} = P$	$Pa_{17} = Q$	$Pa_{18} = R$	$Pa_{19} = S$	$Pa_{20} = T$	$Pa_{21} = U$
Points b_i	1	1	1	4	5	3	6
Weight w_i	0.01	0.01	0.01	0.04	0.05	0.03	0.06
	$Pa_{22} = V$	$Pa_{23} = W$	$Pa_{24} = X$	$Pa_{25} = Y$	$Pa_{26} = Z$		
Points b_i	4	10	4	6	9		
Weight w_i	0.04	0.10	0.04	0.06	0.09		

differentiated expression of subjective preferences than in, e.g., the ranking method. Table 2.2 lists examples of parameters evaluated by the authors according to the scoring scale <1, 10>. Calculation of weights is carried out using Eq. (2.1).

$$W_i = \frac{b_i}{\sum_{Pa=1}^{n} b_{Pa}}, \; i = 1, 2, \ldots, n. \tag{2.1}$$

where: w_i weight assigned to each parameter, b_i—the number of assigned points, n—number of all considered parameters, Pa_i—parameter, i—index of the parameter, b_{Pa}—the total number of points assigned to all parameters.

The resulted weights, determined from experts` assessments, are obvious from Fig. 2.4.

Designated flow rates and state of flood protection structures are identified as the most important parameters related to the assessment of flood protection facilities.

2.3.3 *Creation of Risk Criteria for Risk Factors*

Each parameter (*A–Z*) then has a designated criterion for risk analysis (Table 2.3), divided into five levels. Each level of criterion has a score assigned (0.2, 0.4, 0.6, 0.8 and 1.0), which is entered into the calculation of risk indices.

These risk criteria were defined based on field studies and basic hydrological knowledge and experience as well as resources (Dub and Nemec 1969; Mosný 2002; Zvijáková 2012). Annex A for each parameter Pa_i (*A–Z*) presents the characterization of its determination.

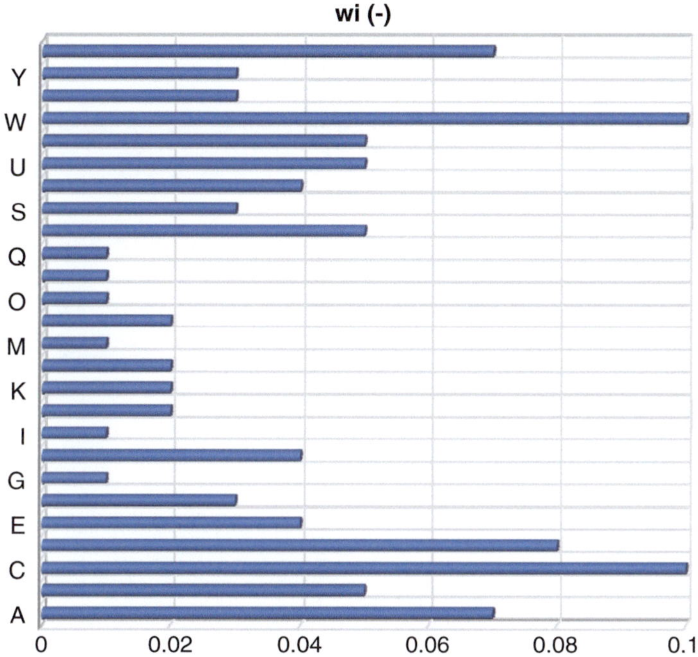

Fig. 2.4 Allocation of weights (w_i) to each parameter Pa_i (*A–Z*)

2.3.4 Determination Score of Criterion (0.2–1.0)

Allocation of scores for each of the proposed parameters is performed in the application of the proposed methodology for a specific proposed activity. The assessor assigns one score SPa_i (0.2, 0.4, 0.6, 0.8 and 1.0) for each parameter Pa_i (*A–Z*) based on Table 2.3.

2.4 Decision-Making

The aim of this step is to determine the average summation risk parameter $ASRP_j$ and average weighted summation risk parameter $AWSRP_j$ for each variant of the activity on the basis of all the allocated scores that reflect environmental impacts of the proposed activity. The risk category of the proposed activity, which determines the size of the risk that the activity poses to the environment, can be classified based on $ASRP_j$ and $AWSRP_j$. By comparing these risk parameters, it is then possible to compare the variants of the proposed construction/activity. To achieve this objective, it is necessary to make some important steps, which are shown in Fig. 2.5.

Table 2.3 Evaluation criteria of risk parameters linked with flood-protection structures

Parameter Pa_i		Score of parameter SPa_i				
		0.2	0.4	0.6	0.8	1.0
A	Maximum specific runoff q_{max} ($m^3 s^{-1} km^{-2}$)	≤ 10	11–50	51–90	91–140	≥ 141
B	100-year discharge Q_{100} ($m^3 s^{-1}$)	≤ 20	21–70	71–120	121–200	≥ 201
C	Design discharge Q_n ($m^3 s^{-1}$)	$\geq Q_{100}$	$<Q_{100}$	$<Q_{50}$	$<Q_{20}$	$\leq Q_5$
D	Average annual precipitation H_z (mm)	≤ 500	501–600	601–700	701–800	>801
E	Forestation *l* (%)	100–80	79–60	59–40	39–20	19–0
F	Coefficient of basin saturation *S* (mm)	≥ 21	16–20	11–15	6–10	≤ 5
G	Character of water course (−)	Stream	Torrent	Middle torrent	Strong torrent	Very strong
H	Average longitudinal slope of the stream i_t (%)	<2	2–5	6–10	11–15	>15
I	Type of the basin (−)	–	Elongated	Transitional	Feathery	–
J	Catchment area S_p (km^2)	≤ 10	11–30	31–60	61–90	≥ 91
K	Soil type (−)	Sandy	Clay-sand, sand-clay	Loam	Clay-loam	Clay
L	Slope of the basin i_s (%)	<2	2–5	6–10	11–15	>15
M	Ecological significance of the area (−)	Very low	Low	High	Very high	Extremely high

(continued)

Table 2.3 (continued)

Parameter Pa_i		Score of parameter SPa_i				
		0.2	0.4	0.6	0.8	1.0
N	Vulnerability of protected species of fauna, flora and their biotopes (n)	0–1	2	3	4	≥ 5
O	Change of the landscape (−)	Preserved harmonic landscape, without disturbing elements (current assessment)	Impact of project on future appearance and character of landscape is not disturbing (preventive assessment)	Impact of project on landscape creates risk of negative impact (preventive assessment)	Presence of symptoms and disturbing elements in landscape (current assessment)	Degradation of landscape
P	Cultural and historical significance of the area (−)	No	Regional	National	National and sup-regional	International and national
Q	Vulnerability of archaeological and paleontological sites and important geological sites (n)	0	1	2	3	≥ 4
R	Permanently resident population in the protected area (n)	<100	101–250	251–500	501–1000	>1000
S	Coefficient of built-up area (−)	≤ 0.02	0.021–0.025	0.026–0.03	0.031–0.035	>0.035
T	Type and importance of transport (point)	≤ 1	2	3	4	≥ 5

(continued)

Table 2.3 (continued)

Parameter Pa_i		Score of parameter SPa_i				
		0.2	0.4	0.6	0.8	1.0
U	Infrastructure of the area (point)	0–1	2–3	4–5	6–7	8
V	Production activity in the area (point)	0–2	4	6	8	10
W	Degree of environmental and human damages (−)	Significant losses of property and human life are not expected	Loss of human life and environmental damage is insignificant	Loss of human life and damage to the environment is unlikely	Loss of human life and damage to the environment is likely	Significant losses of property and human life are expected
X	Total cost of the proposed activity (EUR)	0–100 000	100,001–400,000	400,001–800,000	800,001–1,200,000	>1,200,000
Y	Distance of the place of proposed activity implementation from built-up areas (km)	≥0.501	0.101–0.500	0.051–0.100	0.011–0.050	≤0.010
Z	State of flood protection objects (−)	Construction of a polder and stabilization of the stream	Regulation and stabilization of the stream in an urban zone	Ensure the regulation of runoff water and flow capacity in the stream	Maintenance of the river basin, the river bed and riparian vegetation	No technical flood protection measures are implemented

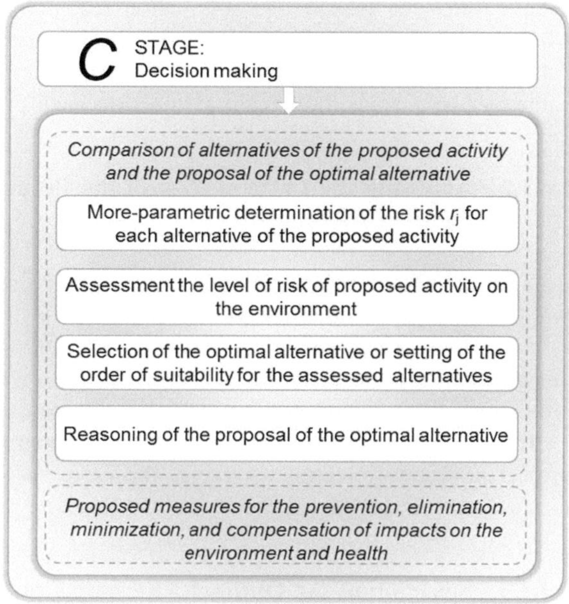

Fig. 2.5 The third stage of impact assessment of the proposed activity

2.4.1 *Comparison of Alternatives of the Proposed Activity and the Proposal of the Optimal Alternative*

For this step, it is necessary to perform the following four tasks.

2.4.1.1 Multiparametric Determination of the Risk r_j for Each Alternative of the Proposed Activity

The project involving construction of flood-protection structures has designated risk indices based on multiparametric risk determination as follows:

- average summation risk parameter $ASRP_j$ is calculated according to:

$$ASRP_j = \frac{\sum_{i=1}^{n} Pa_i}{n} \quad (2.2)$$

- average weighted summation risk parameter $AWSRP_j$ is calculated as follows:

$$AWSRP_j = \frac{\sum_{i=1}^{n} Pa_i w_i}{\sum_{i=1}^{n} w_i} \quad (2.3)$$

where: $ASRP_j$ is the average summation risk parameter of assessed variant (−), $AWSRP_j$ is average weighted summation risk parameter, SPa_i is assigned score (0.2, 0.4, 0.6, 0.8 or 1.0) for each assessed variant (−), n is the number of all considered parameters (−), w_i is the weight assigned to each parameter (−).

2.4.1.2 Assessing the Level of Risk of the Proposed Activity for the Environment

Both the calculated risk parameters of the activity express the level of risk that the proposed activity or assessed variant presents for the environment.

According to the numerical values $ASRP_j$ and $AWSRP_j$, which are calculated using Eqs. (2.1) and (2.2), the different variants of the proposed activity for flood protection are classified into one of the categories according to Table 2.4.

The lower the category of the proposed implementation of water-management project or flood-protection structure, the more risky the activity is for the environment, and the higher is the level of risk of the proposed activity.

2.4.1.3 Selection of the Optimal Alternative or Setting of the Order of Suitability for the Assessed Alternatives

The order of suitability of assessed alternatives of the proposed activity is determined by calculating the average summation risk parameter $ASRP_j$ and average weighted summation risk parameter $AWSRP_j$. The ranking reflects what level of risk the assessed variant represents for the environment. This means that the worst case scenario can be implemented, although that is the least suitable in terms of its possible negative effects on the environment.

2.4.1.4 Reasoning of the Proposal of the Optimal Alternative

Determination of the values of the average summation risk parameter $ASRP_j$ and average weighted summation risk parameter $AWSRP_j$ for assessing the category of the proposed activity is directly related to comparing variants of the proposed action. The optimal variant is then identified on the basis of the lowest levels of $ASRP_j$ a $AWSRP_j$. This choice is justified in terms of expected impacts on the environment.

Table 2.4 Categorization of water structures on the basis of multiparametric risk determination

$ASRP_j(-)/AWSRP_j(-)$	Category of the activity	The level of the risk of the proposed activity for the environment
0.2–0.4	IV.	Very low
0.41–0.6	III.	Low
0.61–0.8	II.	Medium
0.81–1	I.	High

2.4.2 Proposed Measures for the Prevention, Elimination, Minimization, and Compensation of Impacts on the Environment and Health

The task of this step is to propose measures to mitigate the adverse effects of the optimal variant of the proposed activity on the environment.

Article 5(3b) of the EIA Directive requires the developer to include in the environmental information "*...a description of the measures envisaged in order to avoid, reduce and, if possible, remedy significant adverse effects*".

According to Annex no. 11 to Law No. 24/2006 Coll., the measures can be divided into:

- territorial planning measures (e.g., need of harmonization with valid territorial planning documentation, recommendation of change and amendment of valid territorial-planning documentation etc.);
- technical measures (e.g., changes in technology, raw materials, the construction timetable, revitalisation of the area, salvage survey);
- technological measures;
- organization and operation measures;
- other measures (e.g., expected induced investments);
- statement concerning the technical and economic feasibility of the measures.

The proposal of measures to mitigate the adverse impacts of the proposed activity on the environment is an integral part of the methodology, as well as post-project analysis conducted within the EIA process.

References

AGIP KCO (2004) Regulatory basis of environmental impact assessment current environmental status. Methodological aspects of environmental and socio-economic impact assessment. Methodological aspects of environmental and socio-economic impact assessment. http://www.unece.org/fileadmin/DAM/env/eia/documents/CentralAsiaGuidelines/Annex%204%20-%20English.pdf

Catchpole S, Moreno R (2012) Environmental risk and natural hazards. In: EIA's IAIA12 conference proceedings' energy future, the role of impact assessment 32nd annual meeting of the international association for impact assessment 27 May–1 June 2012, Centro de Congresso da Alfândega, Porto, Portugal (www.iaia.org)

CEQ (1978) NEPA (The National Environmental Policy Act) 42 U.S.C. 4321 et seq. CEQ, USA

DEAT (2002) Ecological risk assessment, integrated environmental management, information series 6. DEAT, Pretoria

DEAT (2004) Criteria for determining alternatives in EIA, integrated environmental management, information series 11. DEAT, Pretoria

Demidova O, Cherp A (2005) Risk assessment for improved treatment of health considerations in EIA. Environ Impact Assess Rev 25(4):411–429

Dikmen I, Birgonul MT, Han S (2007) Using fuzzy risk assessment to rate cost overrun risk in international construction projects. Int J Project Manage 25(5):494–505

Dub O, Němec J (1969) Hydrology (in Slovak). SNTL, Praha
EC (2001) Guidance on EIA: Scoping. Office for Official Publications of the European Communities, Luxembourg
Glasson J, Therivel R, Chadwick A (1999) Introduction to Environmental Impact Assessment, UCL Press, London
Křupka J, Kašparová M, Máchová R (2012) Decision processes (in Czech). UPa, Pardubice
Kwiatkowski RE (1998) The role of risk assessment and risk management in environmental assessment. Environmetrics 9(5):587–598
Magness TH (1984) NEPA, the regulations of the US Council on Environmental Quality and the heart of EIA. In: Proceedings of the paper presented to the Dutch-US seminar on environmental management, Washington, D.C.
Mosný V (2002) Hydrology. Morphology of river basin and discharges (in Slovak). STU, Bratislava
National Council of the Slovak Republic (2005) Act of Law No. 24/2006 from December 14th 2005 on environmental impact assessment and on amendments to certain acts
Papa MN, Torres DS (2012) Proceedings of the international workshop on flash flood and debris flow risk management in Mediterranean areas. University of Salerno, Italy
Říha J (2001) Environmental impact assessment. Methods for preliminary decision analysis of EIA (in Czech). ČVUT, Praha
Tichý M (2006) Ovládání rizika, analýza a management (Risk control: analysis and management). C. H. Beck, Prague
Zavadskas EK, Turskis Z, Tamošaitienė J (2010) Risk assessment of construction projects. J Civil Eng Manag 16(1):33–46
Zeleňáková M, Gałaś S, Zvijáková L, Šlezingr M (2012) Calculation of risk posed by flood protection measures as part of the environmental impact assessment process. Pol J Environ Stud 21(5A):459–463
Zeleňáková M, Zvijáková L (2012) Proposal of flood protection objects risk calculation (in Slovak). In: SEA/EIA 2012: 2nd international conference, Slovak Environmental Agency, Nízke Tatry, Tále, 29–30 May 2012
Zvijáková L (2012) Proposal of water construction risk calculation (in Slovak). In: Juniorstav 2012. 14th Ph.D. conference, VUT, Brno, 26 Jan 2012
Zvijáková L (2013) The application of risk analysis in the environmental impact assessment (for selected constructions) (in Slovak). Thesis, Technical university of Košice

Chapter 3
Case Study—Flood-Protection Measures at Kružlov, North-Eastern Slovakia

In the European region, floods are the most common hazards, causing extensive damage and losses. Flooding occurred in 50 of the 53 countries in the WHO European region during the last years, with the most severe floods in Romania, Russia, Turkey and the United Kingdom. It is supposed that climate change will cause more rainfall. This may result in more frequent and more intense floods of various types such as local, sudden floods (flash floods); extensive, longer-lasting pluvial and fluvial floods; coastal floods and snowmelt floods. Even in summer, when the frequency of wet days is supposed to decrease, the intensity of extreme rain showers may yet increase. In addition, the frequency of precipitation over several days is supposed to increase. In consequence, if no measures are taken, river flooding is projected to affect 250,000–400,000 additional people per year in Europe by the 2080s, more than doubling from 1961–1990. The populations most severely affected will be those of central Europe and the British Isles (Ciscar 2009; Menne and Murray 2013). During the past 30 years, flooding killed more than 200,000 people and affected more than 2.8 billion others worldwide. During the past 10 years, in the European region, 1000 persons are reported to have been killed by floods and more than 3.4 million affected (Jakubicka et al. 2010; Menne and Murray 2013).

Recent studies on climate change (EC 2009; Pollner et al. 2010; EEA 2012; Kundzewicz et al. 2013) indicated that central Europe will experience higher frequency of extreme flood events, creating greater demands for flood protection structures (FPS). The use of FPS has become very valuable in many urbanized areas; however, poor management decisions in the implementation of these infrastructures may lead to geomorphological, ecological and/or social ramifications (Everard 2004). For instance, in the past, several channelization works in Europe (for the purpose of flood control) brought adverse ecological consequences to many European river systems (Brookes and Gregory 1983). EIA is a necessary step

M. Zeleňáková and L. Zvijáková, *Using Risk Analysis for Flood Protection Assessment*, DOI 10.1007/978-3-319-52150-3_3

during the early planning stages of FPS in order to gain clear insights into the structures' probable impacts with respect to the different components of the total environment. Likewise, the use of appropriate EIA techniques can aid the decision-makers to formulate appropriate actions based on informed decisions in light of project urgency and limited resources, which are common constraints in the developing countries (Shah et al. 2010).

In this chapter, we apply the methodology for environmental impact assessment which was described in the previous section of this book. To exemplify this application, the proposed activity in the village of Kružlov has been selected. The basic description of the village is found in Sect. 3.1. In Sect. 3.2 the methodology *for environmental* impact assessment of *a* flood-protection structure is applied. The basis for this application is the report: Kružlov—flood protection measures in the village, which was developed by Zeleňáková et al. (2011).

3.1 Basic Data About the Nature and Localization of the Proposed Activity

For the needs of this work, the proposed activity is the construction of flood-protection structures which, as already mentioned, are included in the list of activities requiring the assessment of their impact on the environment.

The location of the proposed activity is in the cadastral territory of Kružlov, through which the stream Slatvinec flows. The village of Kružlov (Fig. 3.1) is located in the north-eastern part of Slovakia, in the upper Šariš area, 13 km west of

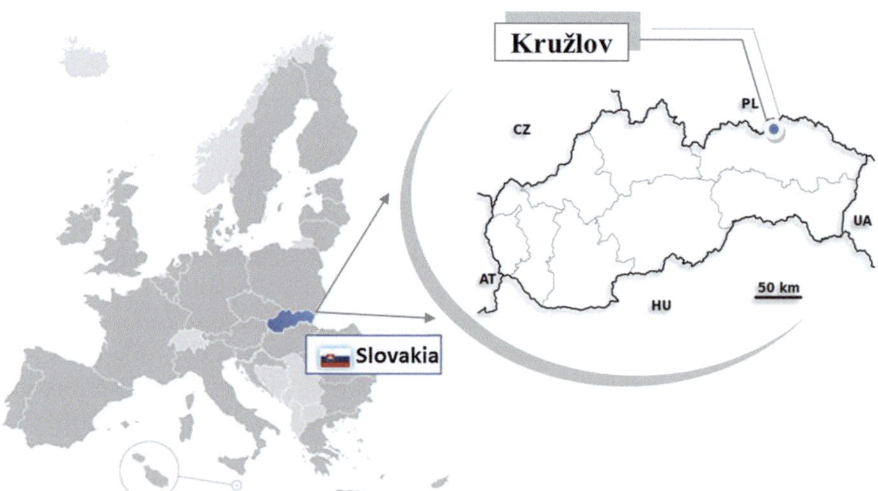

Fig. 3.1 Localization of Kružlov within Slovakia

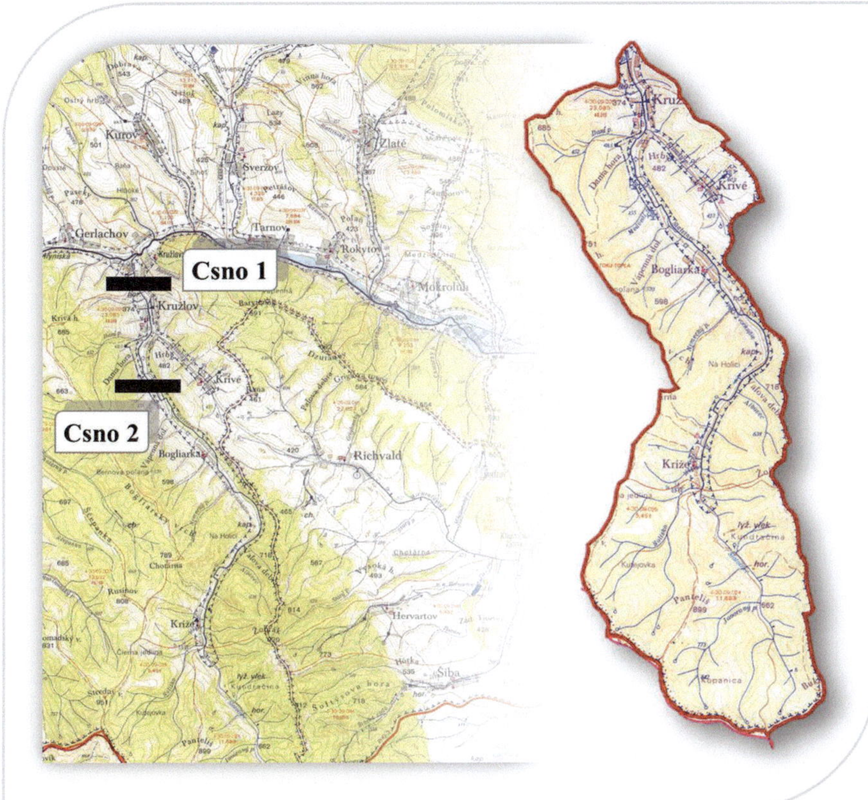

Fig. 3.2 Localization of the activity with close surroundings. *Csno* control station's number

Bardejov. Administratively, Kružlov belongs to the district of Bardejov and the region of Prešov. Figure 3.2 shows the localization of the activity with close surroundings on the scale 1:50,000.

The territory of the activity belongs in the Bodrog river basin and the Topľa sub-basin. The study area with flood-protection structures is located in the cadastral territory of Kružlov and partially extends into the cadastral area of Krivé.

The Slatvinec stream in the village maintains its natural character; the slopes of the stream bed are significantly damaged after floods, and in places, they are very low. There are evident silt deposits in the stream bed. On the right side, there are private lands; along the left side runs the local asphalt road. The stream flows through the village square. Its bed has an irregular trapezoidal cross-section approx. 6.0–13.0 m wide along the bottom. In the area of intended regulation, the stream threatens the existing properties during flood flow, while its erosive activity causes instability of the bank slopes.

3.2 Establishing the Context

3.2.1 Characteristics of the Current State of the Environment in the Affected Area

This part has been prepared according to the report by Zeleňáková et al. (2011).

Characteristics of the Natural Environment including Protected Areas

- *Geomorphologic Conditions*

In terms of geomorphological division, the village of Kružlov is located between three major geomorphological units. The south-western part of the cadastral area is occupied by outliers of the Čergov Hills, which raises the altitude in this section to above 900 m. The north-western part is occupied by the eastern part of the Ľubovnianska uplands with an altitude of around 600 m. The remaining territory belongs to the geomorphological unit of the Ondavská uplands. The altitude in the whole cadastral territory rises from 375 m (confluence of Slatvinec with the river Topľa below the village) up to 905 m (Čergov).

- *Geological Conditions*

The geological structure of the monitored area is relatively monotonous. It consists mainly of the flysch zone of the Outer Western Carpathians. Throughout the territory, there are alternating layers of sandstone and claystone, which date back to the Cretaceous and Paleogene. Quaternary sediments are mainly represented by the deluvial type covering the surroundings of Kružlov near the Topľa river. This geological structure does not predetermine the occurrence of mineral deposits.

- *Climatic Conditions*

From the climatic point of view, the study area is divided into two distinct sections. In the northeast, represented by the rolling relief of the Ondavská hills, the climate is of upland type (slightly warm, humid, and with average annual air temperature of 6–7 °C). The other section is characterized by mountain climate (mildly cold, very wet, and with average annual air temperature of 4–6 °C. The coldest month is January with an average temperature of −5 to −6 °C, and the warmest month is July with an average temperature of 15–17 °C.

The average annual rainfall decreases from north to south in the central part of the area, from the mountainous parts towards the confluence of the Slatvinec stream with the Topľa river. The highest average annual rainfall on the hills of Čergov has values above 900 mm, while the lowest values are just at the confluence of Slatvinec and Topľa, where average annual rainfall is about 600 mm. The area is also rich in snowfall, with snow occurring up to one third of the year in the village and its surroundings. The distribution of major geomorphological units determines the direction of the prevailing winds, with southeast and northwest most frequent compared with the other directions.

- *Soil Conditions*

Most of the cadastral territory of Kružlov is covered with cambisol with low or medium humus content (at a depth of 25 cm). In the central part, there are pseudo soils. Cultivated land is sandy loam, while loamy-sandy soils are covered with forests.

Soils are the basic means of production for agriculture and forestry, and land value plays an important role. In Kružlov soils have low or medium value, as they are quite vulnerable to water and wind erosion. These phenomena cause significant changes to the original cover. The sets of factors that adversely affect the fertility of soils are particularly cool climate, higher humidity, poor mineral substrates for soil formation, lower values of sorption capacity of soils, high soil acidity and segmentation of the relief.

- *Hydrological Conditions*

From the hydrographic point of view, the territory belongs in an area of natural springs. The entire area is part of the Topľa river basin. The principal stream in the area is the Slatvinec. Its source is in the Čergov Hills, on the eastern slopes of Veľká Javorina (1098.7 m) at an altitude of about 940 m above sea level.

Hydrological data were provided by the Slovak Hydrometeorological Institute, Košice Regional Centre, in March 2011 and are listed in Table 3.1.

- *Fauna and Flora*

According to the phytogeographical division (Futák 1980), the study area belongs in the Western Carpathian floral tract (Carpaticum occidentale), the Eastern Beskydy floral territory (Beschidicum orientale), the Eastern Beskydy district and the Čergov subdistrict.

The current vegetation is significantly altered, which resulted from human activity in the past and also at present. Valley plains and accessible slopes have been deforested and turned into meadows, fields, pastures and built-up areas. Despite significant deforested areas in low-lying areas, more than one third of the

Table 3.1 Hydrological data

Stream:	Hydrological number:		
Slatvinec	4—30–09—026		
Profile:	Catchment area:		Distance in km:
Kružlov no. 1	32.20 km^2		2.20 km^2
Maximum flow rate reached or exceeded on average once in:			
20	70	100	years
54.0	70.0	95.0	m^3 s^{-1}
Profile:	Catchment area:		Distance in km:
Kružlov no. 2	40.20 km^2		0.000
Maximum flow rate reached or exceeded on average once in:			
20	70	100	years
60.0	80.0	105.0	m^3 s^{-1}

study area consists of woodland. Of the original vegetation cover, extensive beech stands have been preserved on the northern slopes of Čergov. Reflecting the fact that they occur in hard-to-reach places, they are the best preserved forest communities. Streams are accompanied by grey alder coppices, mixed with common alder and crack willow. These green belts, although incoherent, are of great importance in stabilizing stream banks. Grasslands, occurring at the edges of the forest, occupy one third of the territory.

Spatial differentiation of fauna depends primarily on the differences in vegetation cover. A significant position in the study area is inhabited by animal communities within the forests, particularly those rich in leafy vegetation. Animal communities have arisen in meadows, fields and pastures in deforested areas, and these are strongly influenced by human intervention, especially ploughing, mowing and chemical fertilization. In relation to construction activities and buildings in certain parts of the investigated area, there are also animal communities within human settlements, these having adapted to humans, their facilities and activities.

- *Protected Areas*

There are no protected areas in the cadastral territory of Kružlov. There are no areas included in the territory of the European network of protected areas, which means that there are no European areas of importance, nor protected bird areas. It follows that the cadastral territory of Kružlov, under Law no. 543/2002 Coll. on nature and landscape protection, as amended, valid in the Slovak Republic, is covered by the well-defined dimensions of the fifth and third levels of protection.

Landscape, Landscape Character, Stability, Protection, Scenery

According to the Land Registry, the village territory of Kružlov includes various types of land that form part of the landscape structure and land use: arable land, meadows and pastures, gardens, orchards, forests, water bodies, and built-up areas.

Bardejov District, where the village of Kružlov is located, has an average value of the coefficient of ecological stability of 3.50. This value expresses a quantitative measure of ecological stability/disruption of ecological links in the study area.

General supra-regional TSES was approved by the Slovak government in Resolution no. 312/1992 (defining elements in the scale 1: 200,000) and subsequently transformed into the spatial development plan of the Prešov Region in 2004. There are no elements of supra-regional TSES in the cadastral territory of Kružlov.

Population, its Activities, Infrastructure, and Cultural and Historical Values of the Area

There are 200 houses (181 permanently occupied) and 15 residential units (with 86 apartments total) in Kružlov. There is also a post office and health center (general medicine, dental surgery, and baby clinic). The village territory has an area of 1014 ha and it borders the territories of Gerlachov, Krivé, Bogliarka, Tarnov, Lukov and Richvald. Kružlov currently has 967 residents.

There is a volunteer fire brigade, which was founded in 1926 with 20 members, in Kružlov. It currently has 67 regular members including 13 women. The inhabitants belong to the Greek and the Roman Catholic religions. In the upper part of the village is the Greek Catholic Church of the Virgin Mary, which was built in 1822 of stone and covered with shingles. There are also two chapels: Saint Nicholas' Chapel is next to this church, near the road, and the Chapel of the Ascension of the Lord stands in the cemetery. There is also the Roman Catholic Church of the Immaculate Heart of Mary, built in 1947.

There is an elementary school and kindergarten in the village. Kružlov elementary school is attended by pupils from the surrounding villages—Krivé, Bogliarka, Kríže and Gerlachov. In the school year 2010/2011, there were 111 pupils attending the school. The kindergarten was attended by 42 children. There is a public water-main supply to the village, coming from a surface water source. There is a sewage system built in Kružlov, with no connection to waste-water treatment plants. The village is currently connected to the gas-main supply, but house heating is currently being implemented in a decentralized manner and based on gas and solid fuels (coal, wood, or coke).

Current State of the Quality of the Environment Including Health

According to the environmental regionalization of the Slovak Republic 2010 (MoE and SEA 2010), based on the map "*The grade of environmental quality of the territory*", the village falls within the environmental grade of high quality. According to the "*Regions of Environmental Quality*" map, Kružlov falls within the first grade of environmental quality, which is defined as region of undisturbed environment. According to the "*Ecological quality of cadastral areas according to the structure of land use*" map, the village has the high coefficient of ecological quality of its cadastral area.

In terms of potential noise exposure, the main road no. 77 is significant. While, in the investigated area there is no monitoring of noise, the intensity of noise stress is low in the study area. There are no significant sources of noise.

In the cadastral territory, no natural sources of radiation are recorded, nor extreme anomalies of the Earth's magnetic field. In the study area, distinctly low radon risk has been diagnosed, while occurrences of medium radon risk are insignificant in the eastern part of the village. Prognosis of radon risk is medium.

3.2.2 Explanation of the Reasons Why the Proposed Activity is Required in the Locality

Slatvinec stream, in the village of Kružlov, represents a constant threat of flooding, as Table 3.2 makes clear. In recent years there have been wide-ranging consequences of floods particularly for the environment and the property of residents and the village, as documented in Table 3.3.

Table 3.2 Announcements of the third degree of flood activity in Kružlov (MoE and SWME 2011)

Water stream	village	Years in which the third (worst) degree of flood was announced at least once														Total
		1997	1998	1999	2000	2001	2002	2003	2004	2005	2006	2007	2008	2009	2010	
Slatvinec	Kružlov									X	X		X	X	X	5

Table 3.3 Summary of the causes and consequences of floods (MoE and SWME 2011)

Village	Water stream/section	Year	Brief description of flood	Affected property
Kružlov	Topľa	2006	Rainfall	Damaged left bank of the stream
		2008	Rainfall	Flooded houses, area of the former Piloimpregna Kružľov factory, washing away of state and local roads, damaged road bridge and two footbridges
		2010	Rainfall	Flood damage—not quantified
	Slatvinec	1970	Rainfall	Flooded the state road, gardens
		2008	Rainfall	Flooded waste-water treatment plant, damaged water supply and electricity network, flooded houses and gardens
		2009	Rainfall	Flood damage
		2010	Rainfall	Flood damage

It is therefore necessary to increase the flood protection of the inhabitants and environment in the village. Increasing the flood protection in Kružlov can be achieved by various measures, ranging from less intensive measures of increasing the retention capacity and erosion control of the landscape to highly intensive technical flood-protection structures.

The purpose of the proposed action, construction of a flood-protection structure, is regulation of drainage conditions in order to improve flood protection.

The aim of this paper, based on research results, was to suggest the optimal variant of flood protection that would protect the inhabitants of Kružlov and the surrounding environment from the consequences of torrential rain.

The most appropriate way to protect residential units and economically important sites from the harmful effects of flooding are flood-mitigation measures in the catchment area of the protected site, which will contribute to a reduction in peak flow and distribution of the flood wave over a longer period of time. If such measures are not affected, or are not very effective, it is necessary to direct flood-mitigation measures directly to the protected area or its vicinity. Whereas the time between recognition of the flood hazard and actual flood is short, one of the possible measures is to increase the capacity of the river bed of the watercourse, often in combination with other flood-mitigation measures in the protected site.

Construction of a dry tank (polder) is also an effective flood-mitigation measure, which can achieve a reduction in peak flood flow and redistribution of the volume of flood waves by temporary accumulation of water. After the flood, the polder is emptied and the area may be used in the previously established method or in a similar manner. A polder fulfils technical, environmental and landscaping functions (Water Management Development and Construction joint stock Company 2005).

3.2.3 *Brief Description of Alternatives of the Proposed Activity* A_j (A_0, A_I, A_{II})

Alternative solutions of the proposed activity are essential for the purpose of environmental impact assessment. There are three proposed options.

Alternative 0: The Current State

The so-called zero alternative is the present condition, in which the stream Slatvinec and bridge structures in the village do not have sufficient capacity to dissipate increased flood flows, and which often causes recurring flood situations and consequent damage to the environment and property (see Fig. 3.3).

Alternative I: Stream-Bed Regulation

Variant I refers to proposed adaptations to the stream Slatvinec within the boundaries of the village Kružlov. Stream channelling was in the past and still is today one of the basic construction activities in the country (Šlezingr 2009). The proposed modifications to the flood stream Slatvinec in Kružlov are meant to ensure safe drainage of Q_{100} annual water without subsequent damage to public and private property.

Fig. 3.3 Present state of stream Slatvinec in Kružlov, May 2012

Proposed Solution

The project documentation made by Foraiová and Zeleňáková (2011) includes regulation of the stream *Slatvinec*. The starting point of the proposed regulation (SR km 0.000 = rkm 0.040) is at the edge of the built-up area of Kružlov at the confluence of the Slatvinec and Topľa streams. The end of regulation (ER km 2.208 = rkm 2.248) is upstream of the watercourse, in a southerly direction, beyond the built-up area of the village.

The hydrological data of hundred year return period maximum flow $Q_{100} = 105.0\ m^3\ s^{-1}$, obtained from the Slovak Hydrometeorological Institute (SHMI), were considered for the proposed regulation of the stream.

The proposal involves a partial modification of the stream route, slight reduction of its longitudinal profile and reinforcement of the stream banks. The length of the planned stream regulation is 2.206.0 m, starting at rkm 0.040 and ending at rkm 2.246.

The route is mostly along the original stream bed, adapted to the radius of meanders and lengths of straight sections. The whole area of regulation is designed with 21 curves of radius 45.0–350.0 m.

When planning the stream-bottom levelling, consideration was given to maintaining the stream bed within the original ground, and respecting the safety water level min. 0.40 m above the Q_{100} level along the slope.

From the SR (km 0.000) until km 0.175, a longitudinal bed slope i = 10.0‰ is planned, continuing with a slope of 13.0‰ over a length of 2.051.0 m to ER. The declination of the slope and the achieved stabilization of the bottom of the new channel is accomplished by construction of weirs in the bed.

A trapezoidal shape (Fig. 3.4) of the cross section is proposed with inclination of banks 1:1.5 and bottom width b = 10.0 m; at the ER (outside the built area of the village), the width of the channel narrows to 3.5 m.

In the sections from rkm 0.040 till 0.12580 and rkm 1.730–2.135, the stream bed slopes will be reinforced with planted rockfill of quarry stone of thickness 0.40–0.80 m, height h = 1.90 m, width of bottom 10.0 m and slopes with ratio 1:1.5. The gaps between the stones are to be filled with clay and wattles. The bank reinforcements will be built on stone blocks at the foot of the bank slope.

In the section of rkm 0.12580–1.730, reinforcement of slopes is proposed using stone tiles on cement mortar up to a height of 1.70 m.

The dimensions of the paving stones are at least 300 mm square. Each stone is laid so that gaps are on average 20 mm wide. The tiles will be set into a 150 mm thick substrate-gravel filter layer. Stone tiles with cement mortar will be placed min. 40 cm along the slope above the level of 100-year water. Stone of first class will be used for the reinforcement. The bank reinforcements will be built on precast concrete blocks at the foot of the slope, with dimensions of 90 × 70 × 120 cm, in combination with quarry stone 122.5 × 70 × 60 cm.

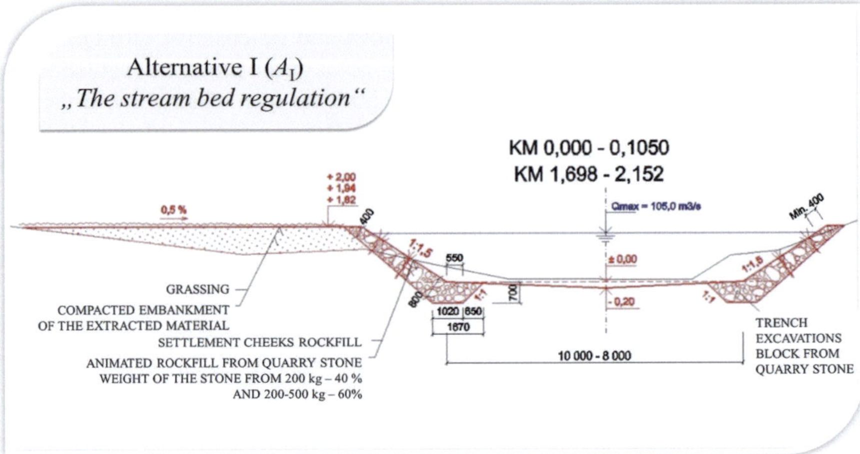

Fig. 3.4 Typical cross section of the stream, scale unspecified (after Zeleňáková et al. 2011)

At the end of the regulation at rkm 2.135–2.248 a trapezoidal shape of river bed is designed with a width of 3.5 m in the bottom and 1:1 inclination of the banks, as the stream passes through a deep notch. The slope of the bed will be reinforced with planted rockfill of quarry stone with thickness of 0.40–0.80 m, height $h = 2.45$ m. The gaps between the stones will be filled with clay and wattles. The bank reinforcement will be built on stone blocks at the foot of the slope.

The slopes above the level of 100-year water will be grassed. The reinforcement of the banks will increase their stability, especially at times of increased water levels. The embankments behind the reinforcement will be made from material extracted from the stream, which will be compacted and replanted.

Over the entire route of regulation 14 weirs are proposed with steps of length 4.0 m, and heights of 0.20, 0.28 and 0.30 m. These weirs will be supplemented by stabilizing weirs at the bottom at a distance of about 20.0–25.0 m along the entire length of regulation. Weirs have dimensions 0.60×1.00 m. The stabilization of the river bed 2.0 m before each weir is made of stone rockfill on a sand-gravel foundation.

- **Alternative II: The Polder Construction**

Variant II considers the proposal of constructing a dry polder, which would be located by the stream Slatvinec above the village. The polder will protect the built area of Kružlov immediately below the dam and will reduce the culmination flow in the lower parts of the stream.

Fig. 3.5 Intended location of the polder

Model Proposed Solution

Project documentation for this variant has not yet been processed. A model proposed solution worked out for the purpose of this work is therefore assumed. Figure 3.5 presents the location of the polder.

The design of the polder is based on the following assumptions:

- the polder will be constructed in the valley of the Slatvinec stream by means of an earth dam;
- the functional structure will be in operation at all times, and its operation is fully automatic, without any required human activity;
- a safety spillway must be designed to safely release flood flows.

The proposed polder will capture the flood wave that occurs when the flow reaches the 100-year maximum—Q_{100}.

The construction of an earth dam creates an accumulation space for the entire volume of the flood wave. The safety spillway is at the height of the water level at which the entire volume of the flood wave is contained. This results from the need for protection against possible failure of the bottom outlet at the time of the flood flow.

Table 3.4 Planned basic parameters of the polder

Stream/village	Volume of the polder V_z till the crest of the spillway (m^3)	Volume V_c at maximum water level on the spillway (m^3)	Flow at hundred-year maximum—Q_{100} (m^3/s)
Slatvinec/Kružlov	50,000–60,000	60,000–70,000	95.0

The polder is designed in a way that after containment of the flood wave the spillway will release any potential further flooding, without threatening the stability of the earth dam. The outlet is sized based on the capacity of the stream bed passing through Kružlov itself, although this flow is achieved only at the maximum level of the polder storage volume. The polder is also provided with a second bottom outlet facility, which will ensure the polder's flood handling in the event of failure of the first one. The basic parameters of the polder for the Slatvinec stream above Kružlov are given in Table 3.4.

Capture of sediments above the construction of the polder is ensured by the weir of the sedimentation tank, situated higher along the stream above the potential flood level. This is one of the most important structures in the whole set, as the catchment of sediments ensures the necessary capacity of the polder to contain the flood wave.

The dam will be constructed with the soil excavated from the area of possible inundation. The safety spillway, for stream regulation about 5 m below the polder and the weir itself, will be made of wired-stone gabions.

3.3 Risk Analysis

The phase of risk analysis consists of stating the parameters, determining the weights of the parameters and creating the requested criteria for the parameters. These steps are described in more detail in Sect. 2.3.

An important step in this phase is the allocation of scores for individual parameters. Table 3.1 presents the values of the individual parameters Pa_i (*A–Z*) specifically designed or calculated (according to Annex A), based on which scores are then assigned for each parameter according to Table 3.5.

Table 3.6 presents the assigned scores SPa_i (0.2, 0.4, 0.6, 0.8 and 1.0) for each parameter Pa_i (*A–Z*) for variants of the proposed activity on the basis of Table 2.3.

From the data above, the risk indices ($ASRP_j$ and $AWSRP_j$) can be calculated for the proposed construction of flood-protection structures in the study area at the village of Kružlov in the northeastern part of Slovakia.

Table 3.5 Determining the value of each parameter for the proposed variants

Parameter Pa_i		Alternative A_i		
		0	I	II
A	Maximum specific runoff q_{max} ($m^3\ s^{-1}\ km^{-2}$)	3.969	3.969	3.969
B	100-year discharge Q_{100} ($m^3\ s^{-1}$)	105.0	105.0	95.0
C	Designated discharge Q_n ($m^3\ s^{-1}$)	70.0	105.0	95.0
D	Average annual precipitation H_z (mm)	600–700	600–700	600–700
E	Forestation 1 (%)	60	60	60
F	Coefficient of basin saturation S (mm)	3.464	3.464	3.464
G	Character of water course (–)	0.237	0.237	0.237
H	Average longitudinal slope of the stream i_t (%)	2.2	1.3	2.2
I	Type of the basin (–)	Elongated	Elongated	Elongated
J	Catchment area S_p (km^2)	23.085	23.085	23.085
K	Soil type (–)	Clay-sand, sand-clay	Clay-sand, sand-clay	Clay-sand, sand-clay
L	Slope of the basin i_s (%)	4995	4995	4995
M	Ecological significance of the area (–)	Low	Low	Low
N	Vulnerability of protected species of fauna, flora and their biotopes (n)	0	0	0
O	Change of the landscape (–)	Presence of symptoms, disturbing elements in the landscape (current assessment)	Impact of project on future appearance and character of the landscape is not disturbing (preventive assessment)	Impact of project on future appearance and character of the landscape is not disturbing (preventive assessment)
P	Cultural and historical significance of the area (–)	International and national	International and national	International and national

(continued)

Table 3.5 (continued)

Parameter Pa_i		Alternative A_i		
		0	I	II
Q	Vulnerability of archaeological and paleontological sites and important geological sites (n)	0	0	0
R	Permanently resident population in the protected area (n)	967	967	967
S	Coefficient of built-up area (–)	0.0381	0.0381	0.0381
T	Type and importance of transport (point)	1	1	1
U	Infrastructure of the area (point)	5	5	5
V	Production activity in the area (point)	4	4	4
W	Degree of environmental and human damage (–)	Loss of human life and damage to the environment is likely	Loss of human life and environmental damage is insignificant	Loss of human life and damage to the environment is unlikely
X	Total cost of the proposed activity (EUR)	0	3,000,000	4,500,000
Y	Distance of the place of proposed activity implementation from built-up areas (km)	0.010	0.050	0.515
Z	State of flood protection structures (–)	No technical flood protection measures are implemented	Regulation and stabilization of the stream in an urban zone	Construction of a polder and stabilization of the stream

3.4 Decision-Making

3.4.1 Comparison of Alternatives of the Proposed Activity and the Proposal of the Optimal Alternative

3.4.1.1 Multiparametric Determination of the Risk R_j for Each Alternative of the Proposed Activity

Table 3.7 shows the calculation according to Eq. (2.2) of the average summation risk parameter $ASRP_j$ and according to Eq. (2.3) of the average weighted-summation risk parameter $AWSRP_j$ for each of the assessed variants of the proposed activity.

Table 3.6 Assigning scores to parameters for proposed variants of flood protection structures

Parameter Pa_i	Alternative A_i			Weight	Alternative A_i		
	0	I	II		0	I	II
	Score of parameter $SPa_i(-)$			$w_i(-)$	$SPa_i\ w_i(-)$		
A	0.2	0.2	0.2	0.07	0.014	0.014	0.014
B	0.6	0.6	0.6	0.05	0.03	0.03	0.03
C	0.4	0.2	0.2	0.1	0.04	0.02	0.02
D	0.6	0.6	0.6	0.08	0.048	0.048	0.048
E	0.4	0.4	0.4	0.04	0.016	0.016	0.016
F	1	1	1	0.03	0.03	0.03	0.03
G	0.4	0.4	0.4	0.01	0.004	0.004	0.004
H	0.4	0.2	0.4	0.04	0.016	0.008	0.016
I	0.4	0.4	0.4	0.01	0.004	0.004	0.004
J	0.4	0.8	0.8	0.02	0.008	0.016	0.016
K	0.4	0.4	0.4	0.02	0.008	0.008	0.008
L	0.4	0.4	0.4	0.02	0.008	0.008	0.008
M	0.4	0.4	0.4	0.01	0.004	0.004	0.004
N	0.2	0.2	0.2	0.02	0.004	0.004	0.004
O	0.8	0.4	0.4	0.01	0.008	0.004	0.004
P	1	1	1	0.01	0.01	0.01	0.01
Q	0.2	0.2	0.2	0.01	0.002	0.002	0.002
R	0.8	0.8	0.8	0.05	0.04	0.04	0.04
S	1	1	1	0.03	0.03	0.03	0.03
T	0.2	0.2	0.2	0.04	0.008	0.008	0.008
U	0.6	0.6	0.6	0.05	0.03	0.03	0.03
V	0.4	0.4	0.4	0.05	0.02	0.02	0.02
W	0.8	0.4	0.6	0.1	0.08	0.04	0.06
X	0.2	1	1	0.03	0.006	0.03	0.03
Y	1	0.8	0.6	0.03	0.03	0.024	0.018
Z	1	0.4	0.2	0.07	0.07	0.028	0.014

Table 3.7 Multiparametric determination of risk for the assessed variants

A_i	0	I	II
$ASRP_j$	0.546	0.515	0.515
$AWSRP_j$	0.568	0.480	0.488

3.4.1.2 Assessment of the Level of Risk of the Proposed Activity for the Environment

Considering the comparison of variants of the proposed activities, it can be stated that:

- Variant 0 "*current state*" in accordance with Table 3.7 is given these values:
  - the average summation risk parameter $ASRP_0$ = 0.546, which in accordance with Table 2.4 represents the IIIrd category of risk to the environment;
  - the average weighted summation risk parameter $AWSRP_0$ = 0.568, which in accordance with Table 2.4 represents the IIIrd category of risk to the environment.
- Variant I "*Water stream regulation*" in accordance with Table 3.7 is given these values:
  - the average summation parameter risk $ASRP_I$ = 0.515, which in accordance with Table 2.4 also represents the IIIrd category of risk to the environment;
  - the average weighted summation parameter risk $AWSRP_I$ = 0.480, which in accordance with Table 2.4 also represents the IIIrd category of risk to the environment.
- Variant II "*Polder construction*" in accordance with Table 3.7 is given these values:
  - the average summation risk parameter $ASRP_{II}$ = 0.515, which in accordance with Table 2.4 also represents the IIIrd category of risk to the environment;
  - the average weighted summation risk parameter $AWSRP_{II}$ = 0.488, which in accordance with Table 2.4 also represents the IIIrd category of risk to the environment.

The assessment shows that all considered variants have different values of the average summation risk parameter $ASRP_j$ and the average weighted summation risk parameter $AWSRP_j$, but they all fall into the same category of risk. Based on the risk category we can assess the level of risk of the proposed activity on the environment (see Table 3.8).

According to the numerical values of the average summation risk parameter $ASRP_j$, the variants of the activity according to Table 2.4 can be considered as classified in the IIIrd category, thus representing a low level of risk to the environment.

Table 3.8 Assessment of the level of risk of the proposed activity on the environment in the village of Kružlov

A_i	0	I	II
$ASRP_j$	0.546	0.515	0.515
Category	III.		
Level of risk of the proposed activity on the environment	Low		
$AWSRP_j$	0.568	0.480	0.488
Category	III.		
Level of risk of the proposed activity on the environment	Low		

According to the numerical values of the average weighted summation risk parameter $AWSRP_j$, the variants of the activity according to Table 2.4 can be considered as classified in the IIIrd category, thus representing a low level of the risk to the environment.

3.4.1.3 Selection of the Optimal Alternative or Setting of the Order of Suitability for the Assessed Alternatives

The design of an optimal variant is based on comparison of the levels of risk for the environment of the proposed activities, on the basis of which the suitability of the assessed variants can be prioritized as follows:

- average summation risk parameter—$ASRP_j$:
  1. Alternative I, Alternative II,
  2. Alternative 0;
- average weighted summation risk parameter—$AWSRP_j$:
  1. Alternative I,
  2. Alternative II,
  3. Alternative 0.

The comparison of average summation risk parameters $ASRP_j$ of the assessed variants is shown graphically in Fig. 3.6.

The comparison of average weighted summation risk parameters $AWSRP_j$ of the assessed variants is shown graphically in Fig. 3.7.

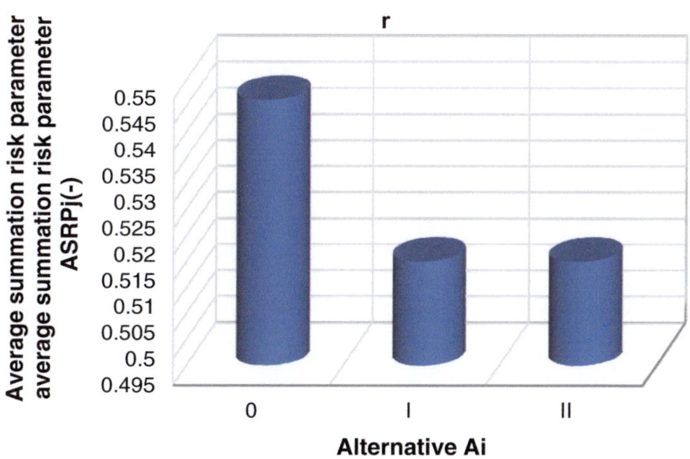

Fig. 3.6 Assessment of the variants of flood-protection structures in the village of Kružlov based on the average summation risk parameter $ASRP_j$

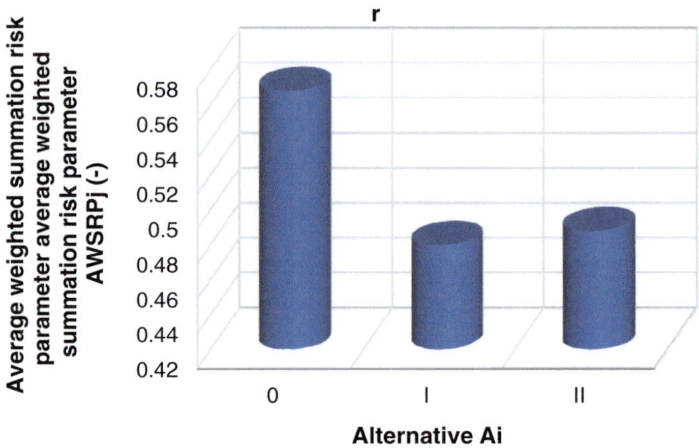

Fig. 3.7 Assessment of the variants of flood-protection structures in the village of Kružlov based on the average weighted summation risk parameter $AWSRP_j$

The lowest value of the level of risk for the environment posed by the proposed activity is the optimal variant. The higher value is a variant which is less acceptable and the highest value of the level of risk is a variant that is least acceptable in terms of the risk for the environment.

3.4.1.4 Reasoning of the Proposal of the Optimal Alternative

On the basis of this assessment, the proposal for the optimal alternative Fig. 3.8 can be justified as follows:

"*The comparison of* variants of the activity—*construction of* flood protection structures *on the stream Slatvinec* at the village *of* Kružlov—shows that *on the basis of the calculated average summation risk parameter* $ASRP_j$ *and the average weighted summation risk parameter* $AWSRP_j$, all three considered variants (*Alternative* 0, *Alternative* I and *Alternative* II) *constitute* different risks for the environment. Based on *the* calculated risk parameters $ASRP_j$ *and* $AWSRP_j$ *it can* be finally appreciated that Alternative I is optimal regarding its potential impacts on the environment, since *it* achieves the lowest value of the calculated risk indices ($ASRP_j = 0.515$ *and* $AWSRP_j = 0.480$); and therefore the variant of "Water stream regulation" *on the* stream Slatvinec in the village *of* Kružlov *is recommended.* This flood protection *structure* is assigned to *the* IIIrd category of water structure, which was designed based on the calculated risk parameters $ASRP_j$ *and* $AWSRP_j$ and represents the overall lowest level of risk for the environment."

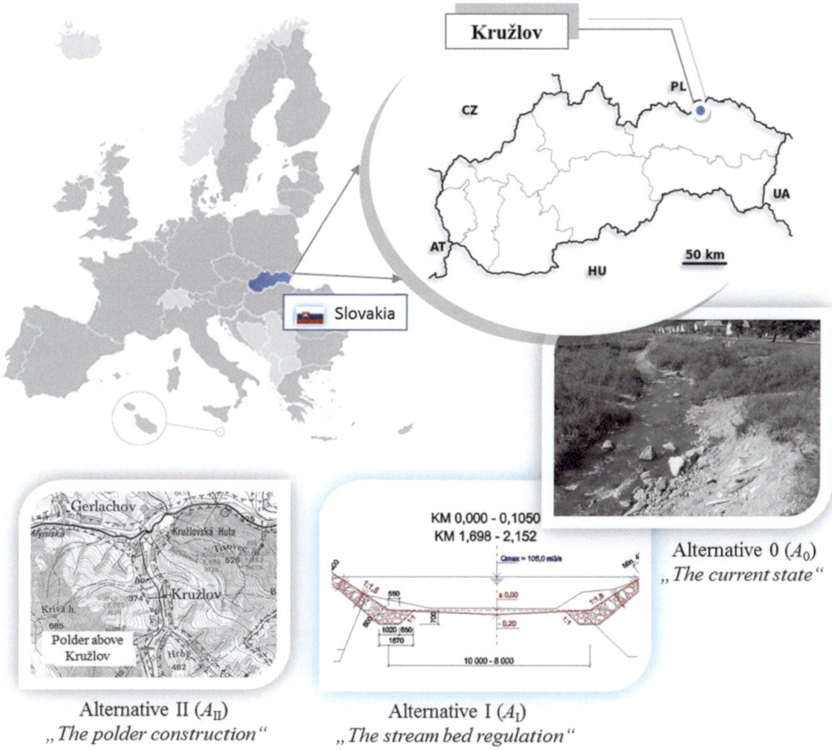

Fig. 3.8 Alternatives for flood protection of Kružlov municipality

3.4.2 Proposed Measures for the Prevention, Elimination, Minimization and Compensation of Impacts on the Environment and Health

Implementation of the plan will be based on the prepared project documentation in accordance with the Planning and Building Act as amended, as well as later regulations determining the subsequent issue of a building permit. Documentation of flood-protection measures will include all the requirements for taking measures to mitigate the negative impacts of the proposed activity on the environment, and to prevent, mitigate, minimize or compensate the expected impacts of the activity which may arise during its implementation. For the implementation of the construction, it is necessary to follow these precautions:

- *territorial planning measures*:
  - the proposed activity with regulations must be approved and incorporated into the land-use planning documentation of the village of Kružlov;

- *technical measures*:
  - adequate replacement with plants of indigenous species must be implemented within the landscaping in suitable locations providing channel flow capacity;
  - the purity of surface and ground water near the stream must be maintained during the implementation of the construction;
  - only the necessary belt of riparian vegetation may be cut before the construction;
  - the felling of trees must be carried out in the non-vegetation and non-breeding season;
  - interventions in the area situated in the immediate vicinity of the construction site must be minimized;
- *technological measures*:
  - the technological part of the work must be carried out in accordance with the prepared project documentation;
- *organization and operation measures*:
  - safety requirements must be respected on the construction site;
  - the necessary extent of the site must be defined;
  - good technical condition of construction machines must be ensured to avoid the penetration of pollutants into the environment;
  - the construction site and dumps of material must be located where there will be no devastation of existing grasslands;
  - devastating interventions by cutting of the existing (although worthless) plants must be minimized;
  - an emergency plan must be drawn up for remedying any ecological damage during the construction;
- *other measures*:
  - the disposal of generated waste must be carried out in accordance with applicable legislation;
  - warehouses, machines and waste construction material must be placed outside the construction site;
  - an appropriate monitoring system of surface and ground-water quality must be maintained during the implementation of the construction;
  - monitoring and evaluation of the activity (post-project analysis) must be performed;
- *statement concerning the technical and economic feasibility of the measures*:
  - agreements with the relevant organizations must be made for disposal of waste generated during the construction;
  - a plan of emergency measures for remedying any environmental damage must be drawn up.

References

Brookes A, Gregory KJ (1983) An assessment of river channelization in England and Wales. Sci Total Environ 153:97–111

Ciscar JC (2009) Climate change impacts in Europe. Final report of the PESETA research project. Publications Office of the European Union, Luxembourg

EEA (2012) Climate change, impacts and vulnerability in Europe 2012. An indicator-based report. Summary. Office for Official Publications of the European Union, Luxembourg

EC (2009) Regions 2020. The climate change challenge for European regions. Brussels. http://ec.europa.eu/regional_policy/sources/docoffic/working/regions2020/pdf/regions2020_climat.pdf

Everard M (2004) Investing in sustainable catchments. Sci Total Environ 324(1–3):1–24

Foraiová K, Zeleňáková M (2011) Project documentation: Kružlov—flood protection measures in the village (in Slovak), July 2011

Futák J (1980) Phytogeographical division of Slovakia (in Slovak). In: Atlas SSR. SAV, Bratislava

Jakubicka T, Vos F, Phalkey R, Marx M (2010) Health impacts of floods in Europe. Data gaps and information needs from a spatial perspective. A MICRODIS report. Centre for Research on the Epidemiology of Disasters, Brussels

Kundzewicz ZW et al (2013) Flood risk and climate change: global and regional perspectives. Hydrol Sci J 59(1):1–28

Menne B, Murray V (2013) Floods in the WHO European Region: health effects and their prevention. WHO, Denmark. http://www.euro.who.int/__data/assets/pdf_file/0020/189020/e96853.pdf

MoE SEA (2010) Environmental regionalisation of the Slovak Republic: 3rd updated and expanded edition (in Slovak). Ministry of Environment/Slovak Environmental Agency, Bratislava/Košice

MoE, SWME (2011) Implementation of Directive 2007/60/EC from 23rd October 2007 on flood risk assessment and management: preliminary flood risk assessment in Bodrog rivr basin, attachements. (in Slovak) http://www.minzp.sk/files/sekcia-vod/bodrog-prilohy.pdf

Pollner J, Kryspin-Watson J, Nieuwejaar S (2010) Disaster risk management and climate change adaptation in Europe and Central Asia. World Bank, Washington D.C. http://www.preventionweb.net/files/15518_gfdrrdrmandccaeca1.pdf

Shah A, Salimullah K, Sha MH, Razaulkah K, Jan IF (2010) Environmental impact assessment (EIA) of infrastructure development projects in developing countries. OIDA Int J Sustain Dev 1(4):47–54

Šlezingr M (2009) Selected reasons of water stream revitalisation (in Czech). In: Infrastruktura. MR Ltd., Brno

WMDC (Water Management Development and Construction joint stock Company) (2005) Catalogue of measures: methodical tool for selection of appropriate measures for solution of main kinds of problems (in Czech). WMDC, Praha

Zeleňáková M, Sarka D, Zvijáková L (2011) Kružlov—flood protection measures in the village. Report worked out according to Act No. 24/2006 Coll. On environmental impact assessment (in Slovak), Košice, July 2011

Chapter 4
Conclusions and Recommendations

At present time of scientific and technological progress, new concepts, tools and methods of solving problems fulfil the principles of a democratic society, which dictate the obligation to provide a healthy environment for future generations.

Part of this challenge is timely assessment of the potential impacts of proposed activities on the environment and human health with acceptable environmental risk. It is necessary therefore to develop theory and apply appropriate methods for the systematic investigation, analysis and evaluation of the effects of projects, constructions, plant, equipment and other activities on the environment and population.

This book is informed by an effort to develop research with the aim of:

- Improving existing qualitative and quantitative methods for assessing the impacts of proposed activities on the environment;
- Achieving a better understanding of relations between probabilities and consequences in determining the risk of each of the considered alternatives;
- Developing methods to stimulate creative approaches in the search for alternatives to proposed activities (constructions) that are environmentally friendly;
- Developing methodologies for the application of assessment of impact of constructions on the environment based on risk analysis.

The book is designed to analyze the current state of the process of environmental impact assessment and to highlight the existing methods used in the assessment process. The work points out the possibility of improving existing methods of assessing the impacts of proposed activities by applying risk analysis in assessing the impact of hydraulic structures on the environment. The use of risk analysis methods for assessing the impact of activities on the environment and human health is undoubtedly an original and innovative proposal. The book has a clearly defined methodology and original concept of solution.

M. Zeleňáková and L. Zvijáková, *Using Risk Analysis for Flood Protection Assessment*, DOI 10.1007/978-3-319-52150-3_4

4.1 Theoretical and Practical Benefits

This book addresses current issues and comparison of alternatives for a proposed activity (construction) and the methods used in the environmental impact assessment of the proposed activity. In this context, it is necessary to develop new approaches based on the application of risk analysis (Zvijáková 2011).

The book will contribute to the strategic goals of sustainable development by proposing research activities that focus on the need to develop a knowledge base as well as developing advanced methods and tools necessary for the process of environmental assessment with the global aim of sustainable resource management, conservation and sustainable use of ecosystems, and ensuring the sustainability of natural and urbanized environments.

The main **theoretical benefits** of the study for environmental impact assessment are:

- The analysis of the current state of theory, practice and effectiveness of the impact assessment process in the world and at the same time pointing out the existing methods used in the EIA process;
- The summary of the EIA process in Slovakia in terms of practice of professionally qualified persons on the basis of a questionnaire survey, which also brought a new opinions on whether the current EIA is compliant, and what in terms of efficiency should be adapted in the current EIA process in the Slovak Republic;
- Analysis of the current state of risk analysis theory and review of risk analysis methods applicable in the EIA process;
- The proposal of the methodology for the EIA process: "*Methodology of the environmental impact assessment of flood protection objects*" using the method of multiparametric expression of risk, which may help to mitigate the negative impacts of proposed activities on the environment by choosing the optimal alternative of the activity.

The proposed methodology can be seen as an important scientific tool, which highlights the quality of the human environment for the further development of society. Other benefits of the work are listed below.

The proposed methodological procedure of environmental impact assessment of flood protection structures consists of:

- proposing indicators A–Z, which are related to the design of flood-protection structures;
- assigning a standardized weight to each of the indicators *A–Z* to determine their significance;
- calculating (based on the method of multiparametric expression of risk) an *average summation risk parameter* $ASRP_j$ and *average weighted summation risk parameter* $AWSRP_j$ assessed for each variant of the construction (flood protection structure), on the basis of which is determined the significance of the risk for the environment.

The proposed methodology assesses the magnitude of the impact on the environment of water structures and activities in the field of water management, according to the purposes of Slovakian Law no. 24/2006 Coll. (National Council of the Slovak Republic 2005) as amended, as well as the European Directive 2014/52/EU of the EIA (Official Journal of the European Union 2014). Similarly, they can be used to evaluate and prioritize risks in areas of the proposed activities.

The main **practical benefits** for EIA are the following. The output of the work—the methodology of impact assessment of water structures on human health and the environment—can be used by the staff of the *Ministry for the Environment* (Section of Environmental Assessment and Management, Department of Environmental Assessment); *District Environment Offices*, and also stakeholders involved in the EIA process. These are mainly *the professionally qualified persons* under the Ministry for the Environment who have professional competence for environmental impact assessment, then *the developers* and *the interested public.*

The application of the proposed methodological approach in terms of the study area along the stream Slatvinec in the village of Kružlov presents results of the assessment in a comparative way, based on graphical-analytical processing of the conclusions of the evaluation of impacts of different flood-protection structures on the environment.

The results of the work will contribute to increasing the effectiveness of the EIA process in practice, not only in Slovakia but also in Europe and worldwide.

4.2 Recommendations for Future Research

During the processing of this study, we have encountered many and various suggestions for further research and development in the area of EIA. This section identifies possible future research, continuing the work presented in this book.

To complete the functional purpose of the environmental impact assessment process, it is necessary to focus attention on the development, updating and publishing of different methodologies for assessing impacts of various proposed activities that would enable better and more objective evaluation of environmental impacts. It may be advisable therefore to investigate:

- Application of other methods of risk analysis in the EIA process and the creation of methodologies for different areas and different types of proposed activities;
- Preparation of a terminological dictionary of technical expressions for the purposes of risk-analysis application in environmental impact assessment;
- Development of new techniques and methodologies, such as:
  - proposal of a methodology of process management using a comprehensive modeling and evaluation tool and appropriate information technology to support the acceleration and automation of the process of environmental impact assessment;

- development of legislation for integrated environmental-health-safety assessment;
- development of a methodology for assessing impacts on Natura 2000 sites and methodologies for assessing impacts on habitats;
- proposal of a methodology for assessing the health risks of noise pollution in the environment;
- creation of a control mechanism for project monitoring;
- preparation of guidelines for the selection of suitable variants (optimal variant) in the process of environmental impact assessment of proposed activities in the form, e.g., "A catalogue of the variants of the proposed activities".

In general, for the existing as well as for the proposed methodologies, some enhancements can be recommended that include:

- Using information technology and software: for complex modeling integrating different modeling standards (e.g., modeling at the level of strategy, objectives, risks, and processes); for statistical evaluation and subsequent prediction of the impacts of activities on the environment;
- Using geographic-information systems for predicting and evaluating the impacts of alternatives of proposed activities.

These proposals can contribute to objectification, standardization and improvement of efficiency of the EIA process. All these areas would contribute to the expansion of scientific approaches in environmental impact assessment.

References

National Council of the Slovak Republic (2005) Act of Law No. 24/2006 from December 14th 2005 on Environmental Impact Assessment and on amendments to certain acts

Official Journal of the European Union (2014) Directive 2014/52/EU of the European Parliament and of the Council of 16 April 2014 amending Directive 2011/92/EU on the assessment of the effects of certain public and private projects on the environment. Text with EEA relevance

Zvijáková L (2011) Environmental impact assessment and risk analysis. In: Ph.D. seminar 2011 of Building and Environmental Institute: 3rd seminar: Proceeding of scientific works. 9–10 February 2011, TU, Košice

Appendix
Description of Individual Parameters

The following is a brief description of each parameter, its possible determination or prediction:

A—Maximal Specific Runoff—q_{max}(m^3 s^{-1} km^{-2})

In Slovakia, for the calculation of runoff from the catchment area, the most commonly used tool is the regional equation according to Dub and Nemec (1969), which gives the maximal specific runoff (A.1):

$$q_{\max} = \frac{A_0}{\left(S_p + 1\right)^{n_0}} \left(1 \pm o_1 \pm o_2\right) \quad \text{(A.1)}$$

q_{max}—maximum specific runoff ($m^3.s^{-1}.km^{-2}$), A_0, n_0—coefficients which express the contribution of a particular territory in Slovakia to runoff (−), S_p—the catchment area (km^2), o_1—coefficient which express the influence of forest cover on runoff conditions (−), o_2—coefficient which express the influence of the shape of the basin on runoff conditions (−).

The individual coefficients can be determined on the basis of the literature, such as Dub and Nemec (1969), Mosný (2002) or Jakubis et al. (2005).

B—100-year Maximum Flow—Q_{100}(m^3 s^{-1})

Maximum flow with a return period of 100 years is the flow in the stream which may be reached or exceeded on average once in one hundred years. The value of flow with a return period of 100 years for each profile is determined by the Hydrological Service of the Slovak Hydrometeorological Institute on the basis of hydrometric observations.

C—Designated Flow—Q_n(m^3 s^{-1})

Flood-protection structures are proposed according to hydrodynamic calculation with respect to the designated flow. The value of the designated flow rate depends on the value in the area which is going to be protected. For the protection of a rural area (extravilan), the designated flow Q_1–Q_5 (return period from 1 to 5 years) is used, and exceptionally Q_{20}, while a built area (intravilan) is protected from the flow Q_{50}–Q_{100} (return period from 50 to 100 years) (Vlasák and Seidl 2010).

M. Zeleňáková and L. Zvijáková, *Using Risk Analysis for Flood Protection Assessment*, DOI 10.1007/978-3-319-52150-3

Table A.1 Designated flow for the implementation of flood-protection structures (Švecová and Zeleňáková 2005)

Type of land use	Q_n for river bed capacity
Continuous built area, industrial area, significant roads and railways	Q_{50}–Q_{100}
Very valuable land, e.g. vineyards, hop fields	Q_{20}
Arable land	Q_5–Q_{20}
Meadows and forests	Q_2–Q_5

The designated flow in a channel is based on the required protection of adjacent lands, which corresponds to their use. Account is also taken of the impact of the proposed action on the flow in the river basin. According to the recommendations (STN 75 2102: 2003; Švecová and Zeleňáková 2005), depending on the nature of land use adjacent to the channel of water flow, the level of proposed flood control should be determined according to Table A.1.

D—Average Annual Precipitation—H_z(mm)

Annual precipitation is the amount of water in liquid and solid form which falls on the horizontal plane at the location in one year. It is expressed as the height of the water column in mm. Figures for this indicator for the study area can be obtained from the Slovak Hydrometeorological Institute; for the Slovak Republic it can be read from the map in Fig. A.1 (produced by the SHI) that is available online as the Atlas of the Landscape of SR (MoE and SEA 2002).

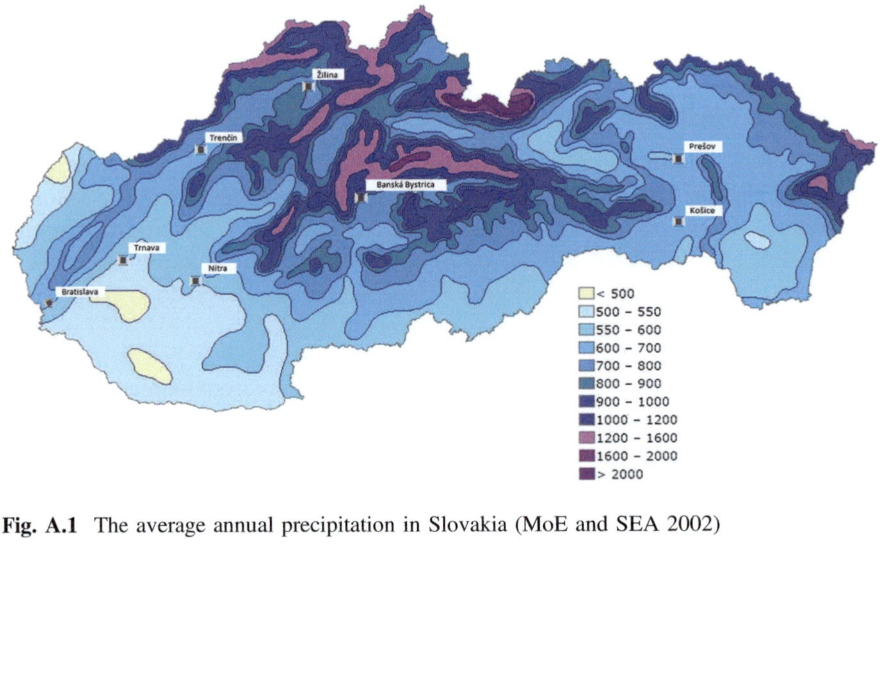

Fig. A.1 The average annual precipitation in Slovakia (MoE and SEA 2002)

***E*—Forestation—*l*(%)**

Forestation l means the proportion of forested area to the total catchment area, expressed as a percentage. It can be determined using formula (A.2):

$$l = \frac{S_L}{S_p} 100 \tag{A.2}$$

where: l–forest cover (%), S_p—catchment area (km^2), S_L—forested catchment area (km^2).

For the Slovak Republic, it is possible to determine forest cover on the basis of the publication Hydrological Conditions issued by The Staff of the Hydrological Service of HMI (1965). The value is reported in whole tens of percent, so that e.g., 0% presents values from 0 to 4, 10% presents values from 5 to 14, 20 … from 15 to 24%, etc.

***F*—Coefficient of Basin Saturation—*S*(mm)**

The saturation capacity of a river basin is an important factor in the origin of floods. This coefficient is calculated using Eq. (A.3) (Jakubis et al. 2005):

$$S = 254 \frac{100 - CN}{10CN} \tag{A.3}$$

where: *CN*—number characterizing the catchment area with regard to water retention (−), *S*—the maximum potential retention capacity of the basin ≈ (mm).

The *CN* value is determined for various types of land use, e.g., from the table *of Medium values of CN curves corresponding to individual hydrologic groups of soil* presented in Jakubis et al. (2005).

***G*—Character of the Water Course (−)**

In addressing a proposal for flood protection-structures, questions come into focus as to which category, in terms of the flow—critical flow and specific energy—in which the stream is classified. To address these issues, the criteria of supercritical flow can be used (Škopek 1984). The criterion for the classification of the flow, as a whole or in its parts, is defined as the coefficient of supercritical flow K_b. Its value is calculated using Eq. (A.4) (Jakubis et al. 2005):

$$K_b = \frac{qOV_sPE\sqrt{S_p + 1}}{L\sqrt{S_L + 1}} \tag{A.4}$$

where: ρ—density of river network (km^{-1}), *O*—length of catchment border (km) V_s—height difference between the source and the mouth of the stream (−), *P*—coefficient of medium soil permeability (−), *E*—coefficient of erosion susceptibility of basin (−), S_p—catchment area (km^2), *L*—length of the stream (km), S_L—forested catchment area (km^2).

Table A.2 Ranking of flows according to the criteria of supercritical flow (Mosný 2002)

Coefficient of supercritical flow	Category
$K_b < 0.1$	I. Water stream
$0.1 < K_b < 0.4$	II. Supercritical flow
$0.4 < K_b < 0.7$	III. Medium supercritical flow
$0.7 < K_b < 1.0$	IV. Strong supercritical flow
$1.0 < K_b$	V. Very strong supercritical flow

According to the coefficient of supercritical flow, a stream is classified into categories according to Table A.2.

The coefficients *P* and *E* characterizing the catchment area, as well as the necessary formulas to calculate other characteristics are published for example in Mosný (2002). Determination of flow according to the criteria of supercritical flow produces one of the suggested indicators—the character of the water course. Flows with a very strong character of critical flow—low depth and high specific energy—are characterized by high-flow velocities and are therefore considered a higher risk of flash floods.

H—Average Longitudinal Slope of the Stream—i_t(%)

The average longitudinal gradient of the stream i_t (bottom or level) is calculated from the relationship between elevation and the length of the stream. It is reported as a percentage or promille (‰) according to the equation (Mosný 2002):

$$i_t = \frac{\Delta H}{L} \tag{A.5}$$

where: i_t—average longitudinal gradient of the stream (%), ΔH—elevation of the stream (m), L—length of the stream (m).

I—Shape of the Basin (−)

The shape of the basin, according to Table A.3, is determined based on the coefficient of the shape of the basin α;

$$\alpha = \frac{S_p}{L^2} \tag{A.6}$$

where: α—coefficient of the shape of the basin (−), S_p—catchment area (km^2), L—length of the stream (km).

Table A.3 The criteria for determining the shape of a river basin (Mosný 2002)

Shape of the basin	$S_p < 50$ km^2	$S_p > 50$ km^2
Elongated	$\alpha < 0.24$	$\alpha < 0,18$
Transitional	$0.24 < \alpha < 0.26$	$0.18 < \alpha < 0.20$
Fanlike	$\alpha > 0.26$	$\alpha > 0.20$

J—Catchment Area—S_p(km^2)

The catchment area of a stream in a certain profile is that part of the basin bounded by the watershed relating to the profile in question. The value of this indicator is determined from water management maps or the publication by the Slovak Hydrometeorological Institute (The Staff of the Hydrological Service of HMI 1965), or calculated using a digital planimeter from a map in scale 1:10,000, or obtained by digitization in geographic software.

K—Soil Type (−)

According to the percentages of particular grain fractions in soil, it is possible to distinguish soil types. Several national and international classifications have been compiled for this purpose. For expressing the soil granularity in Slovakia, the Novak classification is most often used. This classifies soils into seven types according the content of coarse clay (fraction below 0.01 mm). The advantage of this classification is good clarity for users in practice. It allows quite accurate classification of soil types directly during field investigations. The categorization of the types of soil according to the content of particles <0.01 mm and representation of the types of soil in agricultural land in Slovakia are shown in Fig. A.2 (Soil Portal 2013).

L—Slope of the Basin—i_s(−)

The average slope of a river basin is characterized by basin-slope conditions and can be directly quantified as follows:

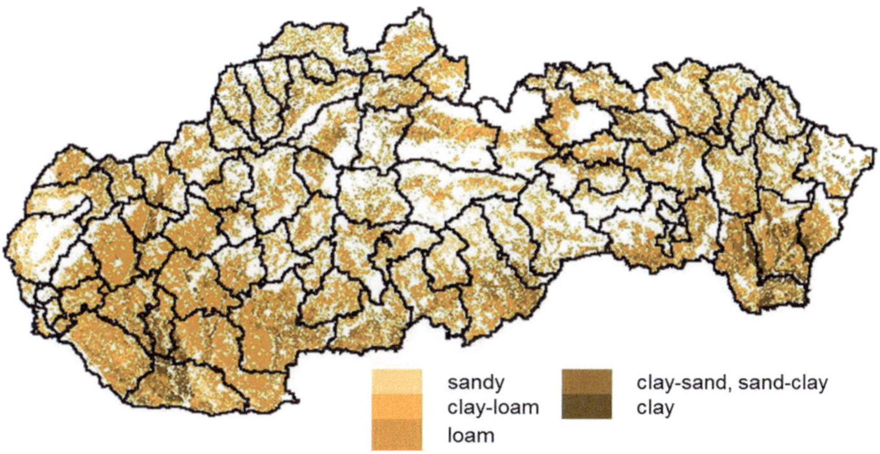

Fig. A.2 Representation of soil types (Soil Portal 2013)

$$i_s = \frac{\Delta H}{S}\left(\frac{L_0}{2} + \sum_{i=1}^{n} L_{v_i} + \frac{L_n}{2}\right) \quad (A.7)$$

where: L_v—length of the *i*th contour (m), ΔH—height difference between contour lines (m), S_p—catchment area (km^2).

M—Ecological Significance of the Area (−)

The ecological significance of the area depends on the presence of protected areas, elements of TSES and other eco-stabilizing elements in the landscape. Eco-stabilizing elements are considered to be forests, shrub vegetation, heterogeneous agricultural areas and various grasslands that are not counted as protected areas, or as elements of TSES. They can be characterized in their mutual combinations; e.g., the territory of Slovakia can be characterized into 9 categories: from the most ecologically significant areas to ecologically insignificant areas (MoE and SEA 2002). The ecological significance of the area, according to the map in Fig. A.3, can be divided into the following classes:

- very small (area with the first degree of protection)
- small (area with the second degree of protection)
- large (area with the third degree of protection)
- very large (area with the fourth degree of protection)
- extremely large (area with the fifth degree of protection).

N—Vulnerability of Protected Species of Fauna, Flora and Biotopes (Number)

The effects of the proposed activity on plant structure and the occurrence of particular species need to be considered objectively. This indicator therefore reflects

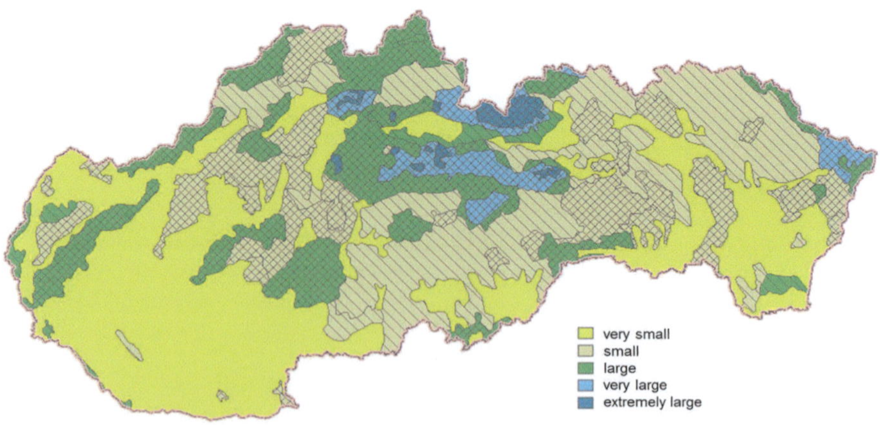

Fig. A.3 The ecological significance of the area (MoE and SEA 2002)

the number of protected plant and animal species that may be affected by the proposed activity.

O—Change in the Landscape (−)

The assessment of impacts on the landscape is based in principle on four states: (i) current state without faults, (ii) current state with failures, (iii) planned activities without faults, (iv) planned activities with the risk of failure (Jančura et al. 2010). This indicator is therefore based on the following criteria:

- *current impact assessment*:
  - preserved harmonious landscape, without interference,
  - presence of symptoms, disturbing elements in the landscape;
- *preventive impact assessment* plan for the future appearance and character of the landscape:
  - without interference,
  - risk of negative interference.

P—Cultural and Historical Significance of the Area (−)

The cultural and historical importance of the area is evaluated according to the occurrence of selected combinations of phenomena, surfaces and objects that are registered in the Central List of Cultural Heritage, as well as the occurrence of sites not included in this list (e.g., a compact area of rural settlements, areas with significant historic residential and solitary dominants, historical infrastructure, and preserved landscape) (MoE and SEA 2002). According to these criteria, the districts in Slovakia have been classified into four levels of cultural and historical significance:

- international and national,
- national and supra-regional,
- national,
- regional.

The geographical representation of this classification for the study area is found in Fig. A.4, available online as the Atlas of the Landscape of SR (MoE and SEA 2002).

Q—Vulnerability of Archaeological and Paleontological Sites and Important Geological Sites (Number)

It is important to know the number of affected archaeological and paleontological sites and important geological sites. This value is determined by the occurrence of these sites in the study area.

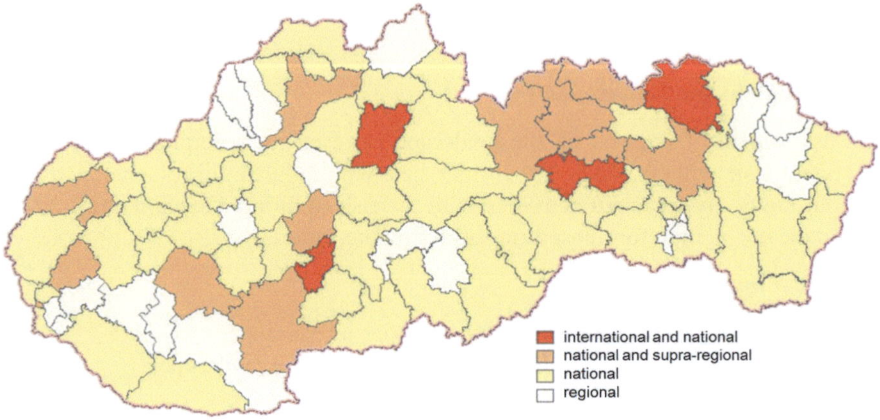

Fig. A.4 Cultural and historical significance of each district of SR (MoE and SEA 2002)

R—Permanent Residents in the Area (Number)

This value represents the number of people who reside in the study area; they may be at risk from flooding.

S—Coefficient of Built-up Area (%)

The coefficient of built-up area in a village is the ratio of built-up area to the total area of the village territory, expressed as:

$$KZO = \frac{ZUO}{CV} \tag{A.7}$$

where: *KZO*—coefficient of built area in the study area (−), *ZUO*—built-up area in the village (m^2), *CV*—total area of the village territory (m^2).

The higher the coefficient of built area, the greater the likelihood of damage caused by potential flooding and thus the higher the flood risk.

Input data for calculation of this coefficient is available on the website of the Cadastral Portal (2013). It is also possible to generate an overview of the economic value of land types for the selected cadastral area.

T—Type and Importance of Transport (Points)

The type and significance of local transport is expressed by scoring (point evaluation) and is given by the sum of all detected means of transport in the study area. The point's scores are given in Table A.4.

U—Infrastructure of the Area (Points)

Infrastructure in a municipality is expressed by scoring and is given by the sum of the identified facilities in the study area. Each group in Table A.5 (I.–VIII.) can be

Table A.4 Allocation of points for the indicator: type and importance of local transport

Type and significance of transport	Points
Roads	
• Local • District and regional • Regional and national	1 2 3
Railways	
• Regional • National	3 4

Table A.5 Allocation of point values for groups of objects

Groups and type of buildings	Points
I. Administrative buildings and shops	1
II. Cultural facilities	1
III. Hotels, restaurants, catering and tourist facilities	1
IV. Warehouses	1
V. Educational facilities	1
VI. Sports facilities	1
VII. Hospitals and social care facilities	1
VIII. Other	1

assigned only one point, even though there may be several kinds of facility falling within the group.

Similarly, the more points are awarded in the study area, the higher the flood risk is, because of higher vulnerability in the area.

V—Production Activity in the Area (Points)

Production activity in a municipality is expressed by scoring and is given by the sum of the existing production activities in the study area. Each type of production activity is scored only once, according to Table A.6.

Table A.6 Type of production activity

Production activity	Points
I. Agriculture	2
II. Craftwork	2
III. Industry	2
IV. Services	2
V. Others	2

W—Degree of Environmental and Human Damage

This indicator refers to the protection of the population and the environment against floods, the population's awareness and preparedness for floods, and the perceived likelihood and expectations of damage to the environment and health of the population.

X—Total Cost of the Proposed Activity (EUR)

This amount represents all funds to be spent on building the proposed flood-protection structures, i.e., the construction budget; or if there are no measures proposed, then this fact is also considered.

Y—Distance of the Location of Proposed Activity from Built-up Areas (km)

This criterion represents the potential for damage to existing infrastructure and the health of the population. It is based on the distance of the proposed flood-protection structure from the nearest built-up area (buildings).

Z—State of Flood-protection Structures (−)

This is a simple classification of existing and proposed measures for flood protection into five categories based on the assessor's decision.

References

Cadastral Portal (2013) http://www.katasterportal.sk/kapor/uhdpFormInit.do
Dub O, Němec J (1969) Hydrology (in Slovak). SNTL, Praha
Jakubis M et al (2005) Ecological flood protection in torrential streams in river basins in territory of Zvolen (in Slovak). TUZVO, Zvolen
Jančura P et al (2010) Methodology of identification and evaluation of visual character of the landscape (in Slovak). Ministry of Environment/Slovak Environmental Agency, Bratislava/Banská Bystrica
MoE, SEA (2002) Atlas of the landscape of the Slovak Republic (in Slovak). Ministry of Environment/Slovak Environmental Agency, Bratislava/Banská Bystrica
Mosný V (2002) Hydrology. Morphology of river catchment (in Slovak). STU, Bratislava
Škopek V (1984) Criteria for distinguishing of river and torrential stream according to the standards ČSN 73 6820, ČSN 736823 and ON 48 2506 (in Czech). Vodní hospodářství 11(1):301–304
Soil Portal (2013) Presence of soil types (in Slovak). Informational Service of Research Institute of Soil Protection
STN 75 2102: 2003 Water stream regulation (in Slovak). SÚTN, Bratislava
Švecová A, Zeleňáková M (2005) Water constructions (in Slovak). TUKE, Košice
The Staff of the Hydrological Service of HMI (1965) Hydrological conditions of Czechoslovakian socialistic republic (in Czech). Part I. Hydrometeorological Institute, Praha
Vlasák J, Seidl M (2010) Catalogue of common equipment for soil revitalization (in Czech). ČVUT, Praha

MIX
Papier aus verantwortungsvollen Quellen
Paper from responsible sources
FSC® C105338

If you have any concerns about our products,
you can contact us on
ProductSafety@springernature.com

In case Publisher is established outside the EU,
the EU authorized representative is:
Springer Nature Customer Service Center GmbH
Europaplatz 3, 69115 Heidelberg, Germany

Printed by Libri Plureos GmbH
in Hamburg, Germany